General Animal Surgery and Anaesthesiology

General Animal Surgery and Anaesthesiology

Libiya M Varghese
Donali Gogoi

2026
Agri Horti Press
New Delhi-110 084

ISBN: 978-93-49690-76-9 (Hardbound)

Published By

Agri Horti Press

New Delhi

E-mail: agrihortipress@gmail.com

Digitally Printed at : Replika Press Pvt. Ltd.

PREFACE

General Animal Surgery and Anaesthesiology is written with the aim of providing a foundational and practical guide to the principles, techniques, and advancements in the field of veterinary surgery and anaesthesia. This book is designed for veterinary students, interns, and early-career professionals who seek to build a strong understanding of surgical protocols, perioperative care, and anaesthetic management in animals, both large and small.

Surgical and anaesthetic practices form an essential part of veterinary medical education and clinical service. As veterinary science advances, there is an increasing emphasis on precision, safety, animal welfare, and post-operative recovery. This book strives to integrate core surgical knowledge with contemporary techniques and evidence-based approaches to anaesthesia and analgesia.

The initial chapters provide a thorough grounding in the general principles of asepsis, sterilization, surgical instruments, and basic operating room protocols. These are followed by detailed sections on surgical techniques commonly performed in general practice, including wound management, soft tissue surgeries, orthopedic procedures, and emergency interventions. The anaesthesiology portion of the book covers essential topics such as the classification of anaesthetic agents, pre-medication, induction, maintenance, monitoring, and recovery, with specific protocols tailored to different species.

Special emphasis has been placed on practical aspects—such as case preparation, pain management, and complication avoidance—that are critical for the safety and success of surgical outcomes. Where appropriate, the book also discusses ethical considerations and animal welfare standards related to surgical interventions and anaesthetic use.

This book reflects current best practices and incorporates recent developments in veterinary technology, including modern anaesthesia machines, monitoring devices, and minimally invasive surgical techniques. Diagrams, flowcharts, and case examples are included throughout to support visual learning and clinical application.

We hope that General Animal Surgery and Anaesthesiology serves as a reliable and accessible resource, equipping future veterinarians with the confidence and competence to perform surgery and administer anaesthesia with care, precision, and compassion.

Libiya M Varghese

Donali Gogoi

CONTENTS

1

INTRODUCTION

Veterinary surgery is surgery performed on animals by veterinarians, whereby the procedures fall into three broad categories: orthopaedics (bones, joints, muscles), soft tissue surgery (skin, body cavities, cardiovascular system, GI/urogenital/respiratory tracts), and neurosurgery. Advanced surgical procedures such as joint replacement (total hip, knee and elbow replacement), fracture repair, stabilization of cranial cruciate ligament deficiency, oncologic (cancer) surgery, herniated disc treatment, complicated gastrointestinal or urogenital procedures, kidney transplant, skin grafts, complicated wound management, and minimally invasive procedures (arthroscopy, laparoscopy, thoracoscopy) are performed by veterinary surgeons (as registered in their jurisdiction). Most general practice veterinarians perform routine surgeries such as neuters and minor mass excisions; some also perform additional procedures.

The goal of veterinary surgery may be quite different in pets and in farm animals. In the former, the situation is more close to that with human beings, where the benefit to the patient is the important factor. In the latter, the economic benefit is more important.

SPECIALIZATION IN SURGERY

In the United States, Canada and Europe, veterinary surgery is one of 22 veterinary specialties recognized by the American Veterinary Medical Association respectively the European Board of Veterinary Specialisation. Those wishing to become board certified must undergo a one-year clinical internship program followed by three years of intensive training in a residency program under direct supervision of board certified veterinary surgeons, including performance of a large number of surgical procedures in such categories as abdominal surgery, surgical treatment of angular limb deformities, arthroscopic surgery, surgery of the foot, fracture fixation, ophthalmic surgery, urogenital surgery, and upper respiratory surgery, etc. Once the minimum requirements of training are met, residents are required to pass a rigorous certification examination before being admitted as members (diplomates) of the American College of Veterinary Surgeons or European College of Veterinary Surgeons.

Veterinary Anesthesia

Anesthesia in animals has many similarities to human anesthesia, but some differences as well. Local anesthesia is primarily used for wound closure and removal of small tumors. Lidocaine, mepivacaine, and bupivacaine are the most commonly used local anesthetics used in veterinary medicine. Sedation without general anesthesia is used for more involved procedures. Sedatives commonly used include acepromazine, hydromorphone, midazolam, diazepam, xylazine, and medetomidine. α2 agonists like xylazine and medetomidine are especially useful because they can be reversed, xylazine by yohimbine and medetomidine by atipamezole. Xylazine is approved for use in dogs, cats, horses, deer, and elk in the United States, while medetomidine is only approved for dogs. Most surgeries in ruminants can be performed with regional anesthesia.

General anesthesia is commonly used in animals for major surgery. Animals are often premedicated intravenously or intramuscularly with a sedative, analgesic, and anticholinergic agent (dogs frequently receive buprenorphine and acepromazine). The next step is induction, usually with an intravenous drug. Dogs and cats commonly receive thiopental (no longer allowed in the UK), ketamine with diazepam, tiletamine with zolazepam (usually just in cats), and/or propofol. Alfaxalone is a steroid anaesthetic used in many practices in the UK to induce anaesthesia in cats and sometimes dogs. It is similar in physiological effect but different in composition to the now withdrawn Saffan. Horses commonly receive thiopental and guaifenesin. Following induction, the animal is intubated with an endotracheal tube and maintained on a gas anesthetic. The most common gas anesthetics in use in veterinary medicine are isoflurane, enflurane, and halothane, although desflurane and sevoflurane are becoming more popular due to rapid induction and recovery.

Common Veterinary Surgeries

Elective Procedures

Elective procedures are those performed on a non-emergency basis, and which do not involve immediately life-threatening conditions. These are in contrast to emergency procedures.

Sterilization Surgery

One of the most common elective surgical procedures in animals are those that render animals incapable of reproducing. Neutering in animals describes spaying or castration (also please see castration). To spay (medical term: ovariectomy or ovario-hysterectomy) is to completely remove the ovaries and often the uterus of a female animal. In a dog, this is accomplished through a ventral midline incision into the abdomen. In a cat, this is accomplished either by a ventral midline abdominal incision, or by a flank incision (more common in the UK). With an ovariectomy ligatures are placed on the blood vessels above and below the ovary and the organ is removed. With an ovariohysterectomy, the ligaments of the uterus and ovaries are broken down and the blood vessels are ligated and both organs are removed. The body wall, subcutis, and skin are sutured. To castrate (medical term: orchiectomy) is to remove the testicles of a male animal. Different techniques are used depending on the type of animal, including ligation of the spermatic cord with suture material, placing a rubber band around the cord to restrict blood flow to the testes, or crushing the cord with a specialized instrument like the Burdizzo.

Neutering is usually performed to prevent breeding, prevent unwanted behavior, or decrease risk of future medical problems. Neutering is also performed as an emergency procedure to treat certain reproductive diseases, like pyometra and testicular torsion, and it is used to treat ovarian, uterine, and testicular cancer. It is also recommended in cases of cryptorchidism to prevent torsion and malignant transformation of the testicles. Please see spaying and neutering for more information on the advantages and disadvantages of this procedure.

Laser surgery offers a number of benefits, including reduced risk of infection, less post-operative pain and swelling, reduced bleeding and improved visibility of the surgical field. Better hemostasis and visibility can in some cases minimize the need for anesthesia and/or reduce overall surgical time.

Controversial Elective Animal Procedures

Other common elective surgical procedures in the United States are declawing in cats (onychectomy), ear-cropping in dogs, tail docking in dogs, horses, and dairy cattle, and livestock dehorning in cattle, sheep, and goats. These procedures have been controversial and recently debated among breeders, veterinary organizations, and animal welfare scientists. The controversy is different for each procedure, but issues arise based on reasons to perform the procedure, differing opinions on techniques and methods, perceived long-term benefits in individual animals, and the development of alternatives. Declawing, for example, consists of removal of the distal phalanges using either a scalpel, scissors or laser. It is typically performed to prevent property damage in house cats, but may also be performed in purebred dogs to meet certain show requirements. Some procedures are illegal in some countries (in the UK, declawing is illegal and tail docking is only allowed in working dogs) and face ethical challenges in others.

Dental surgery

Common dental surgical procedures:

- **Horses:** Floating (grinding down) of uneven teeth edges and removal of wolf teeth.
- **Dogs:** Dental prophylaxis is commonly performed to remove tartar and treat periodontal disease. This procedure is usually performed under anesthesia. Other common procedures include extraction of abscessed or broken teeth, extraction of deciduous teeth, root canals, and removal of gingival hyperplasia and epulides.
- **Cats:** Dental prophylaxis as described above for the dog and treatment and extraction of teeth with feline odontoclastic resorptive lesions (FORLs).

Surgical Oncology

In older dogs and cats tumors are a common occurrence, and may involve any or multiple body systems: skin, musculoskeletal, gastrointestinal tract, urogenital tract, reproductive tract, cardiovascular system, spinal cord and peripheral nerves, the spleen and the lining of body cavities. Common skin tumors include lipomas, mast cell tumors, melanomas, squamous cell carcinomas, basal cell carcinomas, fibrosarcomas, and histiocytomas. Skin tumors are removed through either simple excisions or through excisions needing reconstructive plastic surgery. Common oral tumors include melanomas, fibrosarcomas, and squamous cell carcinomas, which are removed with as much surrounding tissue as possible, including parts of the mandible and maxilla. Other types of cancer requiring surgery include osteosarcomas, stomach and intestinal tumors, splenic masses, and urinary bladder tumors.

Ophthalmic Surgery

Common ophthalmic surgeries in animals include:

- Enucleation of the eye to treat glaucoma or eye proptosis.
- Cataract surgery
- Entropion surgery

- Eyelid tumor removal
- Cherry eye surgery
- Exenteration (complete removal) of the orbit, especially for squamous cell carcinoma in the cat and cow.

Orthopedic Surgery

Common orthopedic surgeries in animals include:

- Ruptured cranial cruciate ligament repair
- For hip dysplasia:
  - Femoral head osteotomy
  - Triple pelvic osteotomy
  - Hip replacement
- Leg amputation
- Bone fracture repair
- Arthroscopy
- MPL - medial patellar luxation

Cardiology Surgery

Common cardiology surgeries in animals include:

- **Balloon valvuloplasty:** Procedure used to alleviate pain symptoms from cardiac problems such as pulmonic, mitral, and tricuspid stenosis. The purpose of this procedure is to create a smoother blood flow throughout the body by reducing harmful effects from obstructed heart valves. It is considered a minimally invasive procedure. The procedure involves putting a balloon-like object inside of the animal's heart; the balloon inflates and deflates in order to alleviate pain and increase blood flow. Complications can occur if the animal's body rejects the balloon, forming a life-threatening allergic reaction.
- **Centesis:** Procedure consisting of the removal of fluid from an animal's body in order to manage congestive heart failure. Draining fluid helps prevent tumors from growing around the heart. Animals undergo no anesthesia because little pain occurs during this operation. Catheters and needles of various sizes are punctured into the thoracic cavity. To remove the fluid, veterinarians will use a syringe suction or vacuum. Post-operatively, swelling might occur around the thoracic cavity or appear bruised, but this is normal and is rarely discomforting to the animal.

- **Devive Embolization of Persistent Ductus Arteriosus (PDA):** PDA is caused by an abnormal blood flow from right to left inside the heart. This causes patients to develop left-sided congestive heart failure. In order to treat this condition, coils are placed inside ducts of the heart. The coils embolize into the pulmonary arteries. Both the ACDO device and the procedure itself are relatively inexpensive.

OTHER COMMON PROCEDURES

Caesarean Section

Caesarean sections are commonly performed in dogs, cats, horses, sheep, and cattle. Usually it is done as an emergency surgery due to difficulties in the birthing process. Certain dog breeds such as Bulldogs often need to have this surgery because of the size of the puppy's head relative to the width of the bitch's birth canal.

Surgery for Gastric Dilatation Volvulus (Bloat)

Gastric dilatation volvulus (bloat) is a common condition in dogs in which the stomach fills with gas, and can become torsed. This requires immediate surgical intervention to prevent necrosis of the stomach wall and death of the dog. During surgery, the stomach is deflated and put back into its normal position. A gastropexy may be performed, whereby the stomach is attached to the body wall to prevent this condition from recurring. A splenectomy or partial gastrectomy may also be required.

Cystotomy

A cystotomy is a surgical opening of the urinary bladder. It is commonly performed in dogs and cats to remove bladder stones or tumors.

Wound Repair

Bite wounds from other animals (and rarely humans) are a common occurrence. Wounds from objects that the animal may step on or run into are also common. Usually these wounds are simple lacerations that can be easily cleaned and sutured, sometimes using a local anesthetic. Bite wounds, however, involve compressive and tensile forces in addition to shearing forces, and can cause separation of the skin from the underlying tissue and avulsion of underlying muscles. Deep puncture wounds are especially prone to infection. Deeper wounds are assessed under anesthesia and explored, lavaged, and debrided. Primary wound closure is used if all remaining tissue is healthy and free of contamination. Small puncture wounds may be left open, bandaged, and allowed to heal without surgery. A third alternative is delayed primary closure, which involves bandaging and reevaluation and surgery in three to five days.

Wounds occurring in the udder and teats of cows are more difficult to repair, due to the difficult access and sensitivity of the organ, and because deep anaesthesia may not be applied to cows.

Foreign Body Removal

A variety of non-edible objects are commonly swallowed by dogs, cats, and cattle. These foreign bodies can cause obstruction of the gastrointestinal tract causing severe vomiting and resulting electrolyte imbalances. The stomach (gastrotomy) or intestine (enterotomy) can be surgically opened to remove the foreign body. Necrotic intestine can be removed (enterectomy) and repaired with intestinal anastomosis. Foreign bodies can also be removed by endoscopy, which although requires general anesthesia does not require surgery and significantly decreases recovery time. However, endoscopic foreign body retrieval is anatomically limited to objects lodged in the esophagus, the stomach or the colon. The condition in cattle is known as hardware disease.

VETERINARY ANESTHESIA

Veterinary anesthesia is anesthesia performed on non-human animals by a veterinarian or a Registered Veterinary Technician. Anesthesia is used for a wider range of circumstances in animals than in people, due to animals' inability to cooperate with certain diagnostic or therapeutic procedures. Veterinary anesthesia includes anesthesia of the major species: dogs, cats, horses, cattle, sheep, goats, and pigs, as well as all other animals requiring veterinary care such as birds, pocket pets, and wildlife.

Specialization in Anesthesia

In North America, the American College of Veterinary Anesthesia and Analgesia is one of 22 specialty organizations recognized by the American Veterinary Medical Association. The ACVAA was recognized by the AVMA in 1975, despite attempts by the AVMA to include anesthesia as a subspecialty of surgery or medicine. As of 2016, there are more than 250 diplomates of the ACVAA. To become an ACVAA board-certified Diplomate, veterinarians must have at least one year of clinical practice experience followed by three years of anesthesia residency training under the supervision of ACVAA Diplomates, have accepted for publication a scientific peer-reviewed research article, and passed both a written and clinical competency examination.

In Europe, the European College of Veterinary Anaesthesia and Analgesia (ECVAA) is one of 23 specialty organizations recognized by the European Board of Veterinary Specialisation. As of February 2014, there are 147 ECVA Diplomates.

Anesthesia Technicians

Anesthesia which is supervised by a qualified technician is safer than anesthesia without a technician. In most private veterinary practices, the technician administers and monitors anesthesia with supervision from the attending veterinarian. In many academic institutions, anesthesia technicians are involved in working with and teaching veterinary students as well as supervising anesthetized cases. The Academy of Veterinary Technicians in Anesthesia and Analgesia is a provisional specialty academy of the North American Veterinary Technician Association and is responsible for licensing technicians as being specialized in anesthesia. For a technician to become specialized, they must be a licensed technician in their state, accumulate 6000 hours of work in veterinary medicine (at least 75% of which must be in anesthesia), 40 hours of continuing education related to anesthesia, demonstrate proficiency in anesthesia skills, and pass a comprehensive written examination.

Application in Animals

Anesthesia is required for many surgical procedures which require the patient to be immobile, unaware, and without pain. Furthermore, anesthesia aims to minimize the surgical stress response. In addition, certain diagnostic procedures require anesthesia, notably stomach or airway endoscopy, bone marrow sampling, and occasionally ultrasound. Aggressive animals may require anesthesia in order to handle and perform a physical exam or obtain blood for testing. Exotic animals frequently require anesthesia for simple procedures (such as taking a radiograph or catheter placement) due to lack of domesticity. Animals may require anesthesia for therapeutic procedures, such as urinary catheterization to relieve obstruction, injection into a mass, or removing fluid from the eye to treat glaucoma.

In addition to anesthesia, analgesia is often managed by anesthesiologists or is included in the considerations for anesthesia.

Techniques

Small Animals

Cats and dogs are frequently anesthetized for surgical procedures. Small animals are most often placed under general anesthesia due to the types of procedures typically performed, the small size of the patient, their suitability to general anesthesia, and the greater degree of control. A balanced anesthesia protocol can be used whereby different drugs with different effects are used so that a high dose of just one drug can be avoided. For example, combining a sedative and an opioid will permit less inhalant anesthesia to be used, improving cardiovascular stability. A one-year study in a teaching hospital shows that dogs and cats typically experience a 1 in 9 chance of anesthetic complications, with a 1 in 233 risk of death. A larger-scale study states the risk of death in healthy dogs and cats as 1 in 1849 and 1 in 895 respectively. For sick dogs and cats, it was 1 in 75 and 1 in 71 respectively. For rabbits, the risk was 1 in 137 and 1 in 14 respectively for the healthy and sick groups.

Horses and Ruminants

Many procedures can be performed on the standing horse with heavy sedation alone. Some procedures may require general anesthesia due to the location of surgery (for example, castration). Other procedures in horses require general anesthesia using an inhalant anesthetic. Horses, due to their complex physiology as performance animals, suffer a number of difficulties that can complicate anesthesia. This results in horses having a higher risk of perioperative fatality - approximately 1 in 400. The number of complications related to fractures or myopathies are approximately 32%

Most procedures in ruminants can be performed standing under sedation and/ or local anesthesia. This strategy is manageable due to the types of procedures being performed, the larger size of the patient, the relative difficulty of general anesthesia, and the cost of the procedure versus the product value of the animal.

Exotic Pets

Anesthesia of exotic animals (guinea pigs, rabbits, birds) is challenging and the higher peri-anesthetic mortality in these species compared to dogs and cats, attests to this fact. These animals are challenging to anesthetize for a number of reasons: very little research has been carried out on safe and effective drug doses for specific species; exotic pets often 'hide' the fact that they are sick and by the time owners realize how sick the pet is the course of the disease is far advanced; the unique anatomy and physiology of exotic pets poses challenges for anesthetic management. For example, the anatomy of the respiratory system of birds, Guinea pigs, and reptiles makes it difficult to induce and maintain anesthesia solely with inhalation agents such as isoflurane and sevoflurane. Injectable drugs such as sedatives or tranquilizers are often used in these patients to facilitate induction and maintenance of anesthesia.

Anesthetic Agents

Most anesthetic agents used in human medicine are used in veterinary medicine. Alpha-2 receptor agonist drugs such as xylazine, romifidine, detomidine, and medetomidine, are used frequently in veterinary species (particularly large animal), but are rarely used in people. Guaifenesin is used as a muscle relaxant prior to anesthesia induction in some animals. Propofol is commonly used in small animal anesthesia, however it is rarely used in large animals due to the cost. 'Propoclear' is also commonly used in smaller branches of small animal surgery as it has a longer shelf life once opened than propofol whilst essentially having the same properties. Butorphanol is rarely used in people but is commonly used in all species. Ketamine, used in children for anesthesia, is used extensively in many species to induce anesthesia or cause heavy sedation. Expensive agents, such as etomidate and desflurane are rarely used outside of university hospitals. Different species have different responses to drugs. For example, horses may experience mania with morphine whereas dogs typically become sedated. Rabbits and guinea pigs are well sedated with midazolam, which can occasionally excite dogs and cats. Tricaine-S/Aqualife TMS (MS-222) is commonly used to anesthetize aquatic animals, including fish, amphibians, and other aquatic, cold-blooded animals.

2

HISTORY

The history of veterinary medicine has seen tremendous change over the past 100 years. It's hard for us to remember that there was a time, not that long ago, when cats, dogs, rabbits, and other domestic animals weren't cared for by a veterinarian.

In fact, it wasn't until the early 1900s that pets began to more commonly receive medical care. Horses that were the primary focus of veterinary medical care, though, as they also served a function as transportation.

Fig. 2.1 Medical Treatment of a Horse

The English word 'veterinarian' as defining one who provides medical care to animals, comes from the Latin verb veheri meaning “to draw” (as in "pull") and was first applied to those who cared for “any animal that works with a yoke” – cattle or horses – in ancient Rome (Guthrie, 1). The association of the term “veterinary medicine” with Rome has encouraged a tendency to begin any discussion of the history of the practice either with the Roman physician Galen (l. 129-216 CE), the earlier Greek “Father of Medicine” Hippocrates (l. c. 460 - c. 379 BCE) or the writer Vegetius (l. late 4th or 5th century CE) when, in fact, the practice was already well-established by the time they lived.

It is as impossible to say when or where veterinary medicine began as it is to definitively claim where or when animal husbandry was first established and dogs first domesticated. Most likely, some form of veterinary medicine followed quickly upon the earliest domestication of animals which, at the latest, is usually dated to 12,000-10,000 BCE though most accounts – especially concerning the dog – date this event much earlier.

Greek & Roman Writers Often Referred to as the “Father of Veterinary Medicine” Only Contributed to What had Already been Established

What is possible, however, is to chart a rough evolution of veterinary practice in ancient civilizations such as China, Mesopotamia, Egypt, and India long before it arrived in Greece and Rome from whence it would later be developed throughout Europe. It is almost a certainty that physicians of Asia and the Near East were practicing veterinary medicine long before the written records which attest to such practice were created, but the accounts which do exist make it clear that the Greek and Roman writers frequently associated with the epithet “Father of Veterinary Medicine” actually only contributed to what had already been established.

Enlightenment writers of the 18th century CE had no knowledge of contributions to the field prior to the Greeks and Romans and so, naturally, began their treatments of the subject with those civilizations; scholarship since then has made it clear that writers such as Hippocrates and Vegetius were later contributors, not pioneers, in veterinary medicine.

Chinese Mythological Origin & Practice

As noted, it is not possible to definitively state where veterinary medicine was first practiced but the earliest documented evidence comes from China. One of the most popular myths of ancient China concerns the god Fuxi (also given as Fu-Hsi, Fu-Shi) and his sister-wife Nuwa who created humanity and bestowed upon them the gifts of civilization. Fuxi is known as “the ox-tamer” for his gift of domesticating animals and clear evidence of domestication was already long established by the time Banpo Village was thriving between 4500-3750 BCE.

Fig. 2.1 Fuxi & Nuwa

According to the myth (as told in the work Classic of Mountains and Seas c. 2600 BCE but thought to have existed orally much earlier), Nuwa created each human being by hand but found the process tiring, as it seemed endless, and so established marriage as a means by which humans could reproduce themselves. After that was taken care of, however, it became clear that the creatures could not survive on their own as they had no idea how to take care of, clothe, or feed themselves. Fuxi taught people how to fish, hunt, write, plant and harvest crops, foretell the future through divination and, finally, domesticate the animals of the wild so they would not have to spend so much time hunting for them.

After domesticating animals, Fuxi is said to have taught humans how to care for them. The earliest examples of veterinary medicine in China have to do with the care of cattle and horses. Doctors known as "horse priests" used acupuncture to successfully treat lame or colic horses c. 3000 BCE. Veterinary practice developed from that point to include other animals and the use of medicinal herbs, incantations, and various procedures in the treatment of illness and injury.

Mesopotamian Practice

In Mesopotamia, veterinarians were also already established by 3000 BCE and the practice was also associated with the divine. The goddess of health and healing was Gula (also given as Ninkarrak and Ninisinna, closely associated with dogs and their protective/healing aspects) who instituted the art of medicine with her consort Pabilsag, her sons Damu and Ninazu, and her daughter Gunurra. Of her children, the most influential was Ninazu – associated with serpents (symbols of transformation), healing, and the underworld – whose symbol was the staff of intertwining serpents, later associated with Hippocrates and, today, the symbol of the medical profession.

Terracotta Plaque Dedicated to Gula

The two primary types of doctors in Mesopotamia were the Asu (a medical doctor who treated illness or injury based on observation and physical treatment of symptoms) and the Asipu (what one today would call a "faith healer" who relied on magical incantations, prayers, and herbs) and both of these types could be veterinarians. There was no distinction between the two – one was not thought to be any more effective or legitimate than the other – and so both "natural" and "supernatural" treatment of illness was practiced together.

The first veterinarian known by name is the doctor Urlugaledinna who served under Ur-Ningirsu (r. 2121-2118 BCE), king of Lagash, son and successor of the great king Gudea (2141-2122 BCE). According to Dr. Saadi F. Al-Samarrai:

Urlugaledinna gave a particular attention to an apparatus consisting of two metal handles attached to two twisted cords with two shafts or lamina which bend upwards at their tips representing a kind of forceps used by Sumerian obstetricians in difficult live births. It proves also that surgical instruments were used for opening abscesses and other minor surgical operations [along with] needles and threads for suturing. (129)

Urlugaledinna's cylinder seal – essentially his personal identification – shows this pair of tongs along with Ninazu's staff of the intertwined serpents and he is more closely associated with veterinarian practice than work with humans though no details as to how his practice developed are available. This is fairly common in Mesopotamian texts which often take for granted an audience's knowledge of the subject. As the Orientalist Samuel Noah Kramer points out:

There were also veterinarians known as "the doctor of the oxen" or "the doctor of the donkeys"; but they are only mentioned in the lexical texts and nothing else is known about them as yet. (Sumerians, 99)

Whatever these early veterinarians did, they had established their practices well enough to be able to define animal-related disease and treatment by the time the law code of Eshnunna was written c. 1930 BCE. The Code of Eshnunna identifies rabies, its effects, and sets the fine to be paid by the owner of a rabid dog who bites someone. The Code of Hammurabi (c. 1754 BCE) recognizes veterinarians as a separate class of medical doctor and sets the fees they are to be paid; clearly establishing veterinary care as a respectable profession.

Egyptian Codification

By the time the Code of Hammurabi was engraved at Babylon, veterinarians in Egypt were already long recognized for their skill and had already produced a work on veterinary science, known today as the Kahun Papyri. Dated to the period of the Middle Kingdom of Egypt (2040-1782 BCE), and specifically to the reign of Amenemhat III (c. 1860 - c. 1814 BCE), the Kahun Papyri contains texts on a number of different subjects ranging from festivals to be held to gynecological issues and treatments, to veterinary practice and diagnoses of illness.

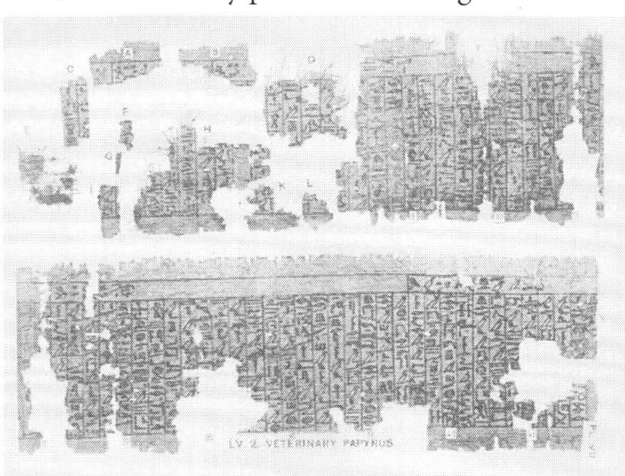

Kahun Veterinary Papyrus

It should come as no surprise that veterinary science would have developed fully in ancient Egypt as the culture placed a high value on animals of all kinds. Although the culture is most closely associated with the veneration of cats, all life was considered sacred, the Egyptian diet was almost wholly vegetarian, and animals were honored through identification with the gods of the land.

Scholar Conni Lord notes how, "as an agricultural society, humans and animals in ancient Egypt would have regularly shared the same space, often to the detriment of each other" (141). This close contact would have necessitated a human response to animal disease in any culture, prompted simply by the instinct of self-preservation, but would have been acted on earlier and more proactively by the Egyptians with their high regard for animals. Some scholars, in fact, have argued that veterinarian practice in Egypt is among the earliest in the world, dating back at least to the time of the Old Kingdom (c. 2613-2181 BCE) if not earlier. Lord comments:

Animals, like their human carers, would have displayed a high incidence of parasitic disease while the harsh glare of the Egyptian sun and frequent dust storms must have led to a high occurrence of both human and animal eye disease. The environmental conditions of ancient Egypt would have facilitated animal and human disease patterns. (142)

Among these diseases was African Trypanosomiasis and, especially, nagana (animal trypanosomiasis) which was spread by the bite of the tsetse fly. Infected flies which bit animals would then spread the disease to humans, causing the "sleeping sickness" which, if left untreated, eventually led to death. The Kahun Papyrus specifically deals with nagana (given as ushau in the text), prescribes remedies, and specifically mentions the importance of washing one's hands before, during, and after treating an infected animal. The Kahun text deals primarily with the treatment of cattle but birds, dogs, and fish – all three kept as pets – are also mentioned.

Indian Advances

Whether Egyptian veterinary science traveled to India or developed there independently is unknown but by the time of the Vedic Period (c. 1500-500 BCE), veterinarians were an established and respected profession in the region. According to scholar R. Somvanshi:

It is believed that the religious priests, who had the responsibility of maintaining cattle, were the first animal healers or veterinarians. A number of Vedic hymns indicate medicinal values of the herbs and it is likely these priests were also apt to [use] their medical knowledge to keep cattle free from ailments. (3)

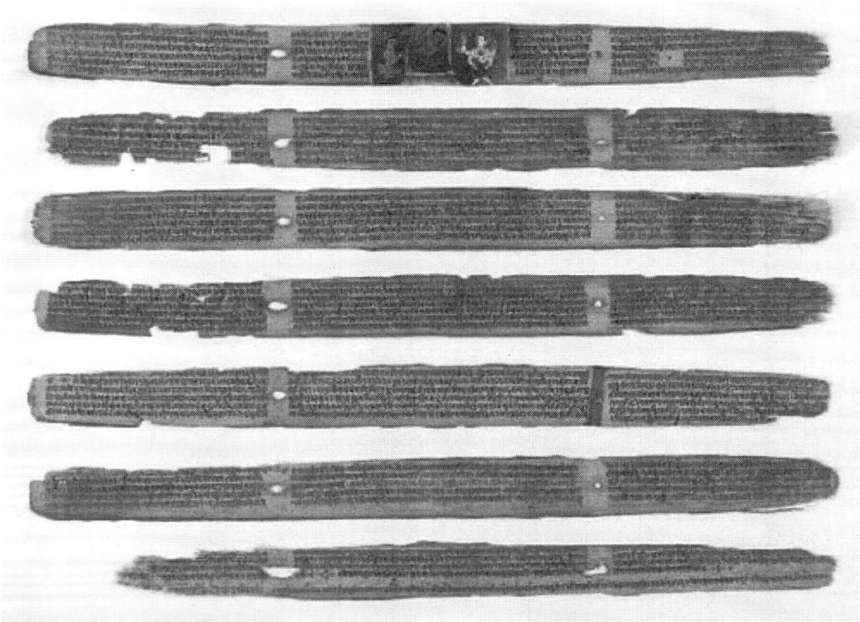

Sushruta Samhita

The great physician Sushruta (l. c. 7th or 6th century BCE), known as the "Father of Indian Medicine" and "Father of Plastic Surgery", developed medical techniques which were used for treating humans as well as animals. His work, Sushruta Samhita (Sushruta's Compendium) is considered the oldest text on plastic surgery in the world, a classic of Ayurvedic Medicine, and the basis for veterinarian practice in India. Somvanshi writes:

Animals received good medical care in ancient India. Physicians treating human beings were also trained in the care of animals. Indian medical treatises like Charaka Samhita, Sushruta Samhita, and Harita Samhita contain chapters or references about care of diseased as well as healthy animals. There were, however, physicians who specialized only in the care of animals or in one class of animals only; the greatest of them was Shalihotra, first known veterinarian of the world and the father of Indian veterinarian sciences. (5)

Shalihotra (l. c. 3rd century BCE) was a physician who dedicated himself solely to the care of animals. His work, the Shalihotra Samhita, dealing with veterinary science, is based on Sushruta's earlier work on human anatomy, physiology, and surgical techniques; these were adapted for the care of animals. By the time of the great king Ashoka (r. c. 268 - c. 232 BCE), the first veterinary hospital in the world was established in India with its underlying vision based on the work of Shalihotra.

GREEK & ROMAN DEVELOPMENTS

Hippocrates Emphasized a Completely Empirical Approach to Diagnosing & Treating Both Humans & Animals

The Greeks follow the same paradigm as other civilizations in, no doubt, developing some form of veterinary science shortly after the domestication of animals but one of the most comprehensive treatments of the subject comes from Hippocrates who emphasized a completely empirical approach to diagnosing and treating both humans and animals. Hippocrates was the first Greek healer to maintain that illness was caused by environmental factors, diet, and lifestyle and was not a punishment from the gods nor an infliction caused by evil spirits or the restless dead. He was not, however, the first in history to make this claim as it was suggested much earlier by the Egyptian polymath Imhotep (l. c. 2667-2600 BCE) and later by Sushruta and Shalihotra in India.

Hippocrates suggested diet as one of the most important aspects of maintaining health, in humans or animals, as well as regular exercise, sunlight, massage, relaxation, and elevation of one's mood, aromatherapy, and regular baths. Although his work focused on human health, it also extended to the welfare of animals. By 130 BCE, a man named Metrodorus of Lamia (in Thessaly) was famous for his skill in healing animals based on Hippocrates' work. He was especially known for his work with horses and was well respected as a veterinary surgeon.

There is no doubt that Greek medical practices were adopted by the Romans and, most famously, by Galen who recognized the similarity in human and animal physiology. He was able to treat his patients as well as he did through his knowledge of anatomy derived from his work with animals. He correctly assumed that what was harmful to an animal would be equally so to a human and, conversely, what would encourage health in the one would most likely do so in the other.

Even so, Galen's work with animals is usually eclipsed by that of the Roman writer Publius Flavius Vegetius Renatus (best known simply as Vegetius) whose Guide to Veterinary Medicine (Digesta Artis Mulomedicinae) became the standard reference work for those working in the veterinary field. Nothing is known of Vegetius outside of his works which deal with diseases and treatment of horses and cattle.

He had clearly read Hippocrates as he insisted on the same basic understanding a veterinarian should have, prior to treating a patient, that disease was naturally occurring and was not caused by divine or supernatural influences. His work has led many through the centuries to consider him, often above other Roman or Greek writers, as the "Father of Veterinary Medicine" for the scope of his work and its influence on the development of veterinary science.

Portrait of Seven Notable Greek Physicians & Botanists

After the fall of the Western Roman Empire in 476 CE, and the rise of Christianity, prior knowledge of veterinary science was lost. Dr. Earl Guthrie comments:

The Church forbade dissection and autopsies and confiscated and destroyed much of the literature on the subject of Veterinary Medicine. During this time no new literature was written. The only work that was done was by the Arabs in Spain. Because of their love of the horse and excellent horsemanship, they were interested in the diseases of the horse.

The lack of interest in veterinary medicine stemmed from the medieval Church's insistence that animals had no immortal soul and so were not worthy of medical treatment. If one's pet cat or dog died, according to the view of the Church, it was of as little consequence as the death of a fly or flea. It was not until the late

12th and early 13th centuries CE that Europeans began to again pay attention to the health of animals as it affected human wellbeing. Even so, this interest was primarily focused on the health of horses and cattle, the one used in warfare and for transport, and the other for food and agricultural endeavors. The health of an animal for its own sake would not become a focus until much later.

It was not until the Age of Enlightenment (c. 1715-1789 CE) that veterinary medicine would again be regarded with any serious interest. Those who wrote on the subject, however, had no knowledge of the contributions of the Chinese, Sumerians, Indians, Egyptians, or others and believed the works of the Greeks and Romans to be the earliest in the field. It was natural, therefore, that Hippocrates, Galen, and Vegetius should be the ones whose work informed the first veterinary schools of Europe.

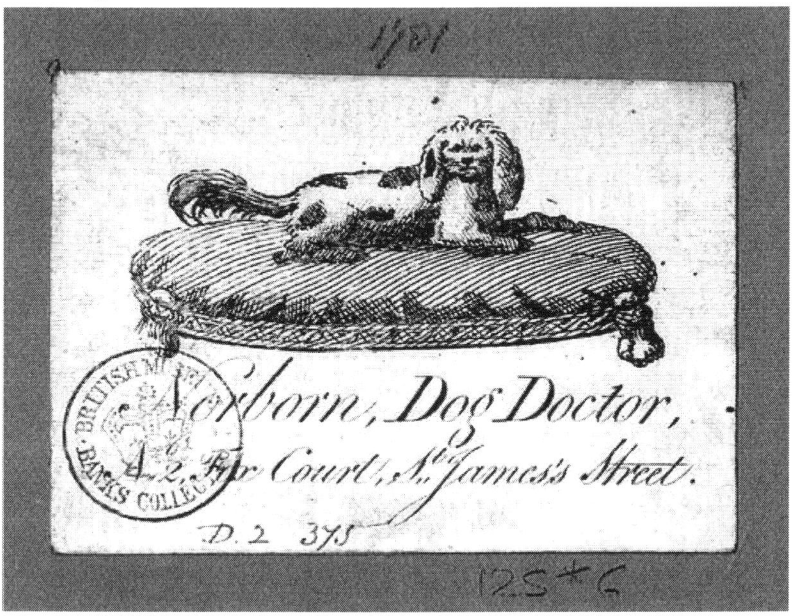

Trade Card of a Veterinary Surgeon

The first veterinary educational institute in Europe was founded in France in 1762 by the veterinary surgeon Claude Bourgelat (l. 1712-1779 CE) as a response to the massive deaths of cattle due to the plague. The students at Bourgelat's school made such impressive strides in research, diagnosis, and treatment, that the French king Louis XV (r. 1715-1774 CE) established a Royal School of veterinarian medicine in 1765 CE and other schools would follow in other European countries through 1791 CE and up through 1862 CE with the first veterinary school in the United States – the Veterinary College of Philadelphia - established in 1852 CE.

Bourgelat is sometimes cited as the "Father of Veterinary Medicine" for founding his school but this claim, still appearing in the present day, ignores the establishment of the veterinary college in India under Ashoka and the work of the Egyptian physicians who created the Kahun text. The most recent "Father of Veterinary Medicine" is the well-known American physician James Harlan Steele (l. 1913-2013 CE) who is, justly, hailed for elevating public awareness of the care and safety of animals.

Although Dr. Steele's accomplishments in the field are admirable, he – like other Western physicians before him referenced as "firsts" in the discipline – is by no means the first. The actual "Father of Veterinary Medicine" – or "mother" for that matter – may never be known but the practice has a much longer, and grander, history than is generally assumed.

The Early Roots of Veterinary Medicine

The first known veterinary practice came into being in 9,000 BC in the Middle East. Sheep herders used rudimentary medical skills to treat their animals, which included the dogs that watched over their herds. Thousands of years later, in Egypt between 4,000 – 3,000 BC, medical treatment of animals became more common, but was still largely undeveloped. Ancient humans began domesticating cats, fowl, and dogs, and their owners considered them as members of their household–like many of us do today.

In approximately 1,900 BC, someone captured the first written accounts of veterinary medicine in four sacred Hindu texts. Within these texts, two distinct writings outlined the fields of human and animal medicine. Millennia later, in 1850, archaeologists discovered fragments of an ancient veterinary medical textbook made of papyrus. This text covers diseases relating to birds, cattle, dogs, and fish. Horses were the primary focus of ancient medical care as they were economically important for transportation, agriculture, and trade.

The Last Few Centuries of Veterinary History

In the 1760s, Claude Bourgelat established the first school of veterinary medicine in Lyon, France. Popular modern thought is that this was the founding of veterinary medicine, despite some level of animal medicine predating 9,000 BC. With the opening of the school in France, the scientific study of veterinary medicine was officially born.

As human medicine flourished and progressed through the last few centuries, so did veterinary medicine. In the 1700 and 1800s, we discovered treatments for cholera, Typhoid fever, and tuberculosis. We could then apply those treatments to protecting farm animals from these same deadly diseases. In Britain, the field of veterinary medicine had its roots in the Odiham Agricultural Society. The Society was the first to apply scientific principles to the treatment of animals. From there, another institution was born, the London Veterinary College, in 1791.

In America in 1863, the American Veterinary Medical Association came into being as a way to promote the field. Its purpose was to oversee the advancement of veterinary medicine and its practitioners. The Food and Drug Administration (FDA) added a Veterinary Medical Branch in 1965 to oversee veterinary pharmaceuticals. It later became the Center for Veterinary Medicine (CVM). The work of the CVM is still essential in that it oversees regulations relating to food, medicine, and other products for animals.

The Rise of Technology

Despite these advancements, only in the past 30 years has there been a huge shift in veterinary medicine. We've moved away from a focus on medical care only for livestock. Veterinarians opened thousands of animal hospitals devoted to the care of cats and dogs and other small pets. These small, furry creatures have moved into our homes and hearts and have family status.

Advancements in veterinary technology and diagnostics continue to improve our ability to detect diseases early. This motivated regular wellness exams to catch help prevent disease or detect it early. Wellness and preventive care allows us to give our pets a better chance of good health and a higher quality of life. Using tools like digital radiology, advanced diagnostic and surgical equipment, and pharmaceuticals, we are better equipped to successfully treat them and keep them living longer.

Attitudes about animals and their status have also prompted changes in how we view wellness care. It now includes things such as dental cleanings, grooming, and alternative treatments like massage and acupuncture. Thirty years ago, the only veterinary equipment for diagnosis was the x-ray. We now have digital x-ray, ultrasound, MRI, advanced laboratory testing, laparoscopic technology, and more.

Anesthesia and anesthetic monitoring for animals has also made leaps and bounds, making surgery safer and more effective for our pets. With a better understanding of animal pain, we are able to manage pain and discomfort in our pets with new medications and alternative therapies. New treatments for cancer mean we can frequently give pet owners much more time with their pets, and give the pets a good quality of life during those extra months or years.

Along with these medical changes, technology now enables us to communicate with our patients in ways that increase speed, convenience, and efficiency. Texting, online booking, apps–we're much more connected. There is no aspect of veterinary medicine or care that has not changed. The use of new technologies continues to increase the speed of change. Those changes make it possible for us to care for our animal companions with increasing success.

The team at OVRS is amazed to consider the history of veterinary medicine and its ongoing evolution. We are at the forefront of veterinary medicine and take full advantage of medical and technological advances. Our specialists push those boundaries to ensure that our patients have the best care available. If you have any questions relating to our therapies and services, please do not hesitate to call.

3

SURGICAL INSTRUMENT

A surgical instrument is a tool or device for performing specific actions or carrying out desired effects during a surgery or operation, such as modifying biological tissue, or to provide access for viewing it. Over time, many different kinds of surgical instruments and tools have been invented. Some surgical instruments are designed for general use in all sorts of surgeries, while others are designed for only certain specialties or specific procedures. Accordingly, the nomenclature of surgical instruments follows certain patterns, such as a description of the action it performs (for example, scalpel, hemostat), the name of its inventor(s) (for example, the Kocher forceps), or a compound scientific name related to the kind of surgery (for example, a tracheotome is a tool used to perform a tracheotomy).

The expression surgical instrumentation is somewhat interchangeably used with surgical instruments, but its meaning in medical jargon is the activity of providing assistance to a surgeon with the proper handling of surgical instruments during an operation, by a specialized professional, usually a surgical technologist or sometimes a nurse or radiographer.

CLASSIFICATION

There are several classes of surgical instruments:

- Graspers, such as forceps
- Clamps and occluders for blood vessels and other organs (e.g. hemostats)
- Surgical scissors
- Bone cutters: unpowered or powered saws, drills and pliers-like devices

- Needle drivers, also known as needle holders, used to hold suture needle while it is passed through tissue and to grasp suture while instrument knot tying.
- Retractors, used to spread open skin, ribs and other tissue
- Distractors, positioners and stereotactic devices
- Mechanical cutters (scalpels, lancets, trocars, Harmonic scalpel, rongeurs etc.)
- Dilators and specula, for access to narrow passages or incisions
- Suction tips and tubes, for removal of bodily fluids
- Sealing devices, such as surgical staplers
- Irrigation and injection needles, tips and tubes, for introducing fluid
- Powered devices, such as cranial drills and dermatomes
- Scopes and probes, including fiber optic endoscopes and tactile probes
- Carriers and appliers for optical, electronic, and mechanical devices
- Ultrasound tissue disruptors, cryotomes and cutting laser guides
- Measurement devices, such as rulers and calipers

An important relative distinction regarding surgical instruments is the amount of bodily disruption or tissue trauma that their use might cause the patient. Terms relating to this issue are 'atraumatic' and minimally invasive.

INTRODUCTION

There are a wide array of instruments available for use during a surgical procedure, each with their specific uses and advantages (and disadvantages).

As a surgeon, it is important to know not only their names but when they should be used, and each speciality will have their own specific kit.

In this article, we cover the main instruments found in a general set for open abdominal surgery.

Ratcheted Forceps

Allis

Allis forceps are toothed surgical instruments, used to grasp firm tissues (such as fascia).

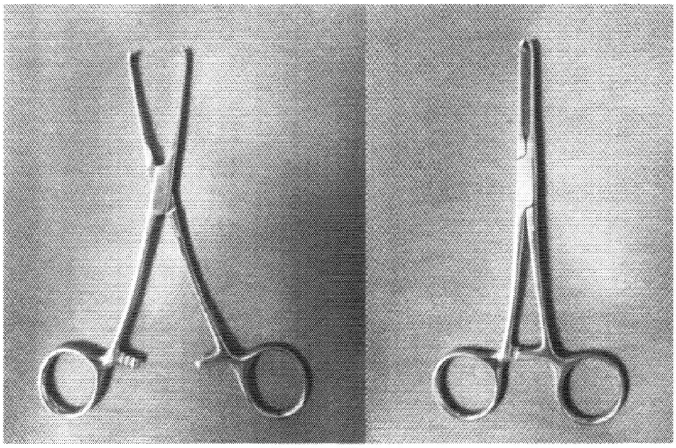

Fig. 3.1(a) Allis Forceps.

Babcock

Babcock forceps are jawed instruments with a smooth end, allowing for more delicate structures (such as bowel) to be held in an atraumatic manner

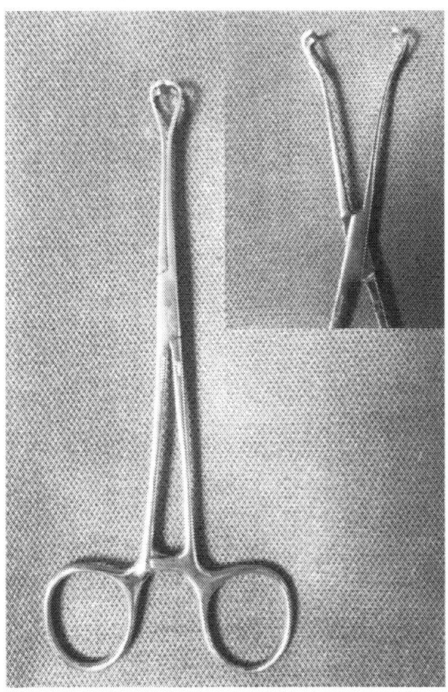

Fig. 3.1(b) Babcock tissue forceps

Dunhills

Dunhill forceps are small forceps with curved serrated ends, often used to grasp vessels prior to ligation

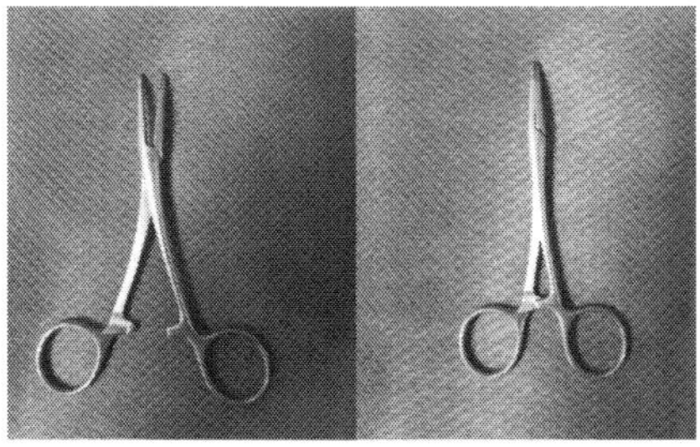

Fig. 3.2 Dunhill forceps

Lane Tissue Forceps

Lane tissue forceps have interlocking teeth, allowing the grasping of tough tissues, such as fascia

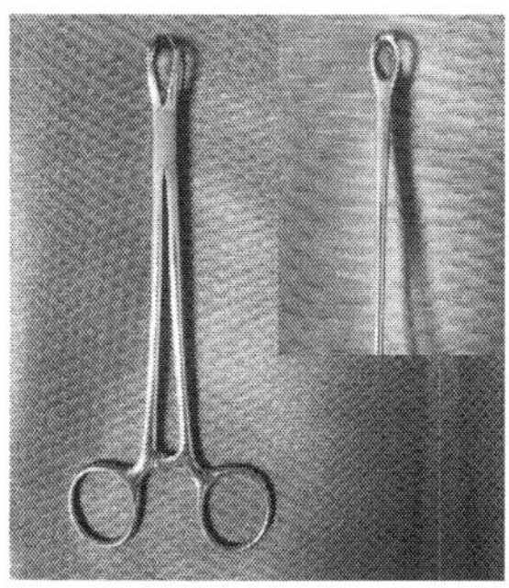

Fig. 3.3 Lanes tissue forceps.

Littlewoods

Littlewood forceps have blunt ended teeth, used to grasp tough tissues, such as fascia. They are often used to gain entry via the umbilicus for laparoscopic surgery.

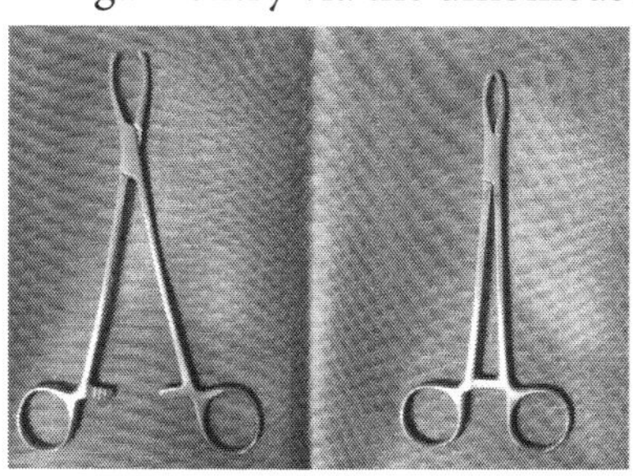

Fig. 3.4 Littlewood forceps

Sawtell

Sawtell forceps are curved forceps with a serrated end, often used to grasp vessels prior to ligation

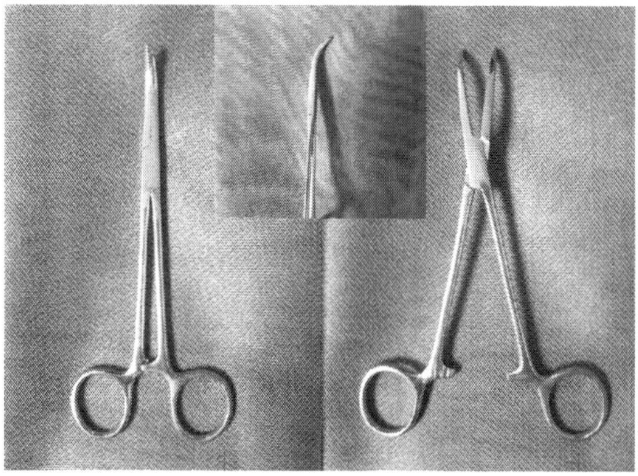

Fig. 3.5 Sawtell forceps

Spencer Wells

Available as curved or straight, Spencer Wells forceps can be used to clamp medium to large sized vessels prior to ligation

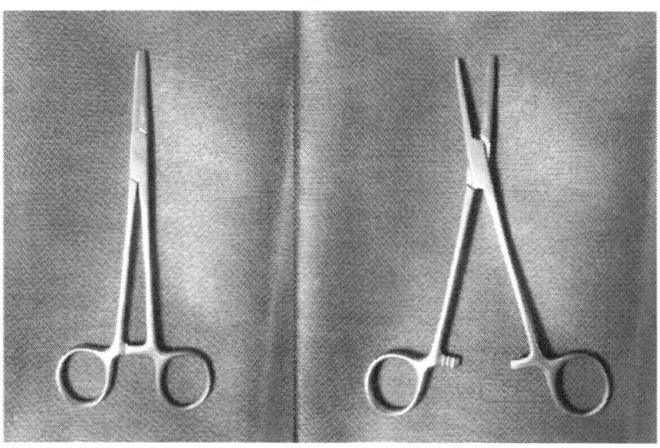

Fig. 3.6 Spencer Wells forceps

Non-Ratcheted Forceps

Debakey

Debakey forceps are non-toothed forceps used in a wide variety of procedures, importantly can be used to grasp tissues (such as bowel) without damaging them

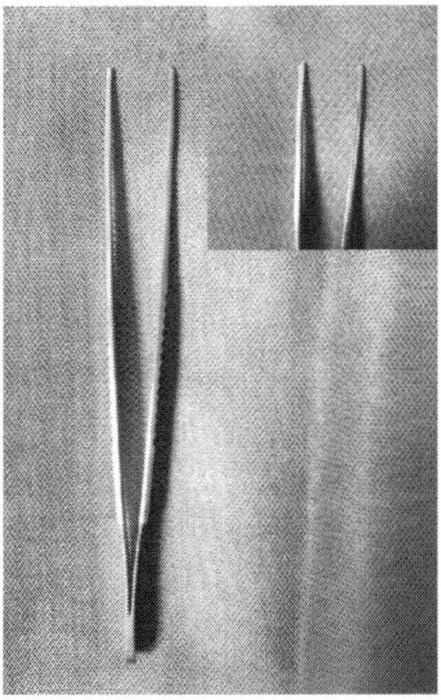

Fig. 3.7 Debakey forceps

Lanes

Lanes forceps are toothed instruments, useful for grasping tissues and are widely used within all surgical specialities (however not to be used to grasp bowel)

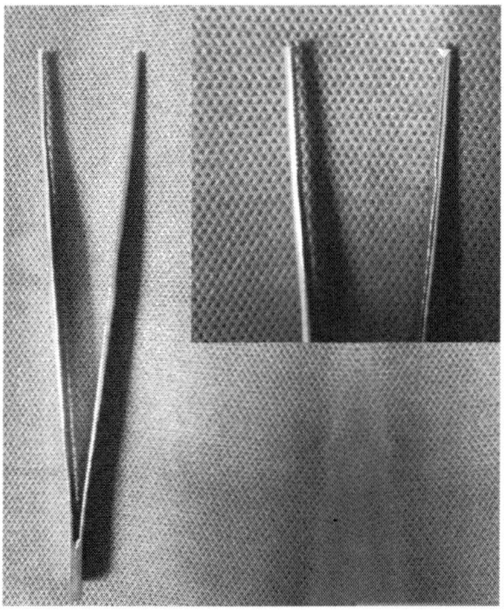

Fig. 3.8 Lanes forceps

Gillies

Gillies forceps are narrow toothed forceps, often used to grasp skin (not to be used to grasp bowel)

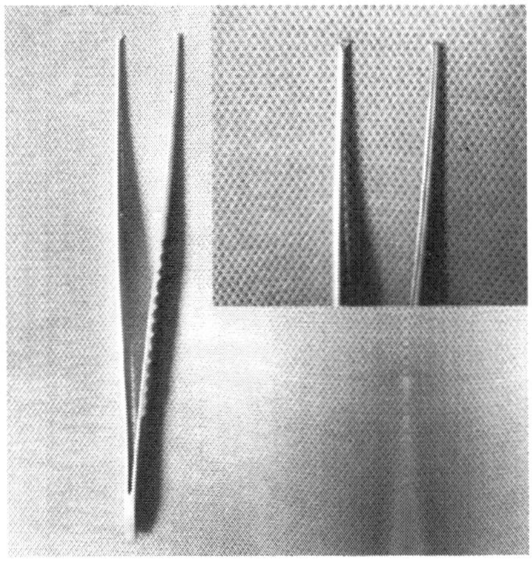

Fig. 3.9 Gillies forceps

SCISSORS

Mayo

Mayo scissors are heavy scissors, with semi-blunt ends, often used to cut thick tissues or sutures, either straight or curved

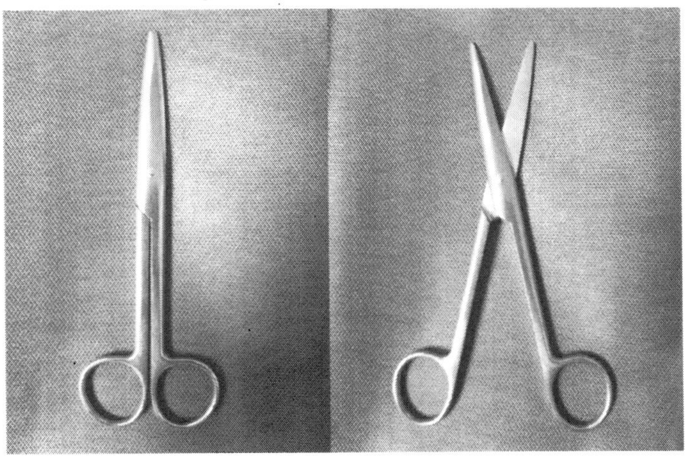

Fig. 3.10 Mayo scissors

McIndoe

McIndoe scissors are used often for cutting or dissecting tissues, characteristically have a curved blade

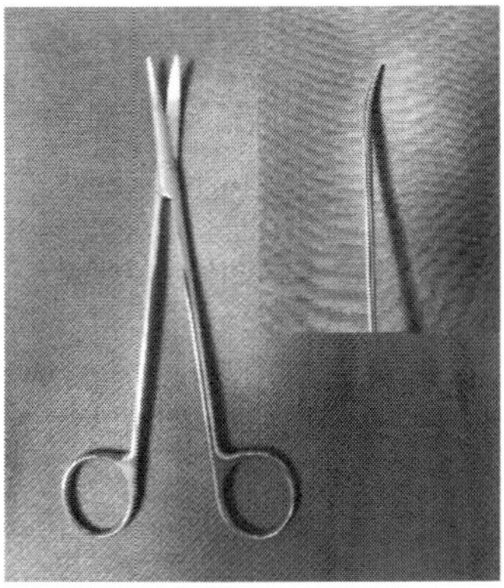

Fig. 3.11 McIndoe scissors

Retractors

There are multiple types of retractors available, all of varying sizes and length, such as Langenbeck, Travers, Morris, or St Mark's retractors. There use is dependent on the operation and wound involved, however the most common are discussed below.

Langenbeck

The Langenbeck retractor is a hook shaped retractor with an L-shaped end, used to keep back tissue or separate the edges of wounds. They come in small, medium, or large sizes, depending on the depth of the cavity required

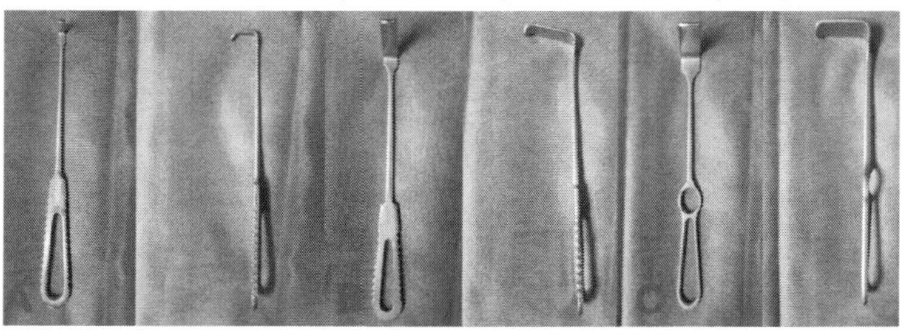

Fig. 3.12 Langenbeck retractors (A) Small (B) Medium (C) Large

NORFOLK AND NORWICH

The Norfolk and Norwich retractor is a self-retaining retractor, used to keep deep wounds and cavities open, with blunt ends to reduce risk of iatrogenic tissue injury

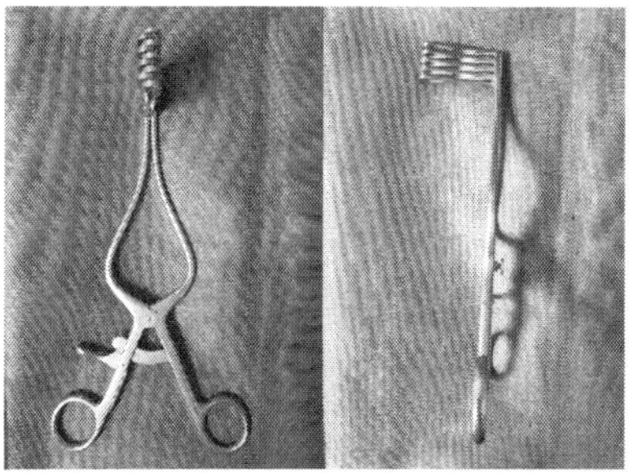

Fig. 3.13 Norfolk and Norwich retractor

Travers

The Travers retractor is a self-retaining retractor, with short ends for use in keeping superficial wounds open

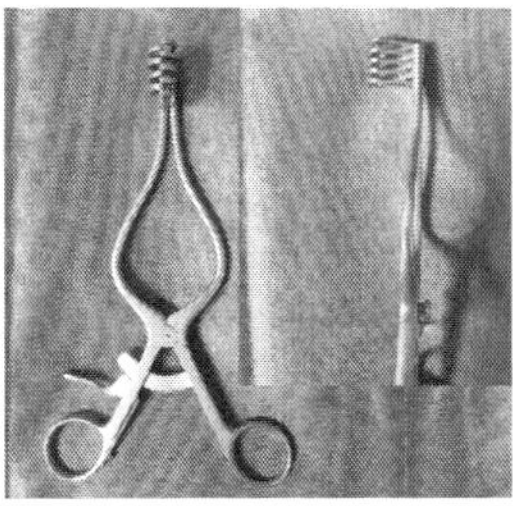

Fig. 3.14 Travers retractor

Other

Needle Holder

For holding needles, will come in varying sizes and length depending on the needle and tissue in question

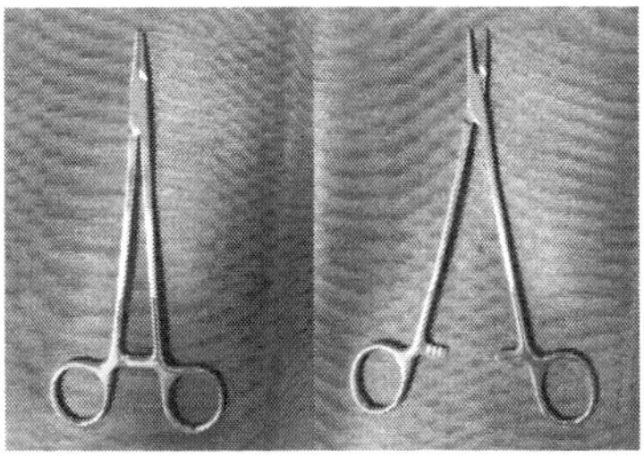

Fig. 3.15 Needle holder

Sponge Holding Rampley

Rampleys can be used for handle guaze or sponges

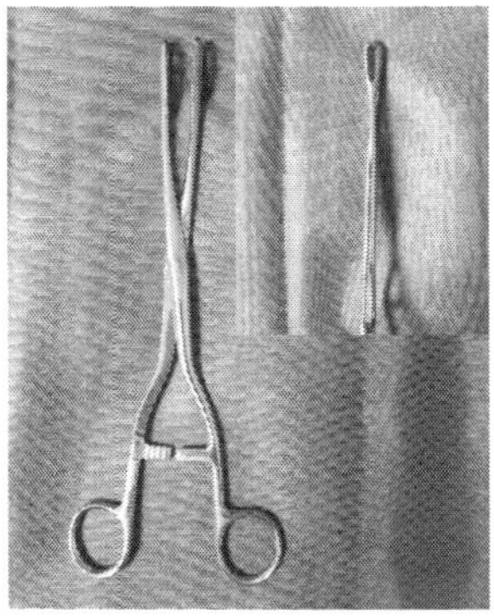

Fig. 3.16 Sponge-holding Rampley

TOWEL CLIPS

Towel clips are most commonly used to secure towels or drapes in place, however can be used to grasp tissue if required

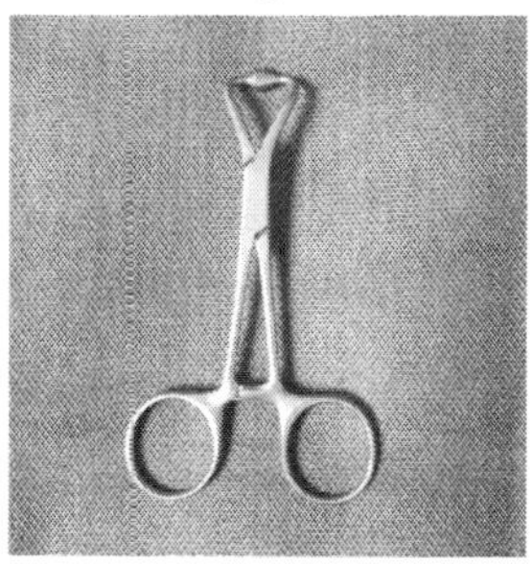

Fig. 3.17 Towel clips

DISCRETIONARY INVASIVE PROCEDURES ON ANIMALS

Numerous procedures performed on domestic animals are usually more invasive than purely cosmetic alterations, but differ from types of veterinary surgery that are performed exclusively for urgent health reasons. Such procedures have been grouped together under the technical term 'mutilatory' by the Royal College of Veterinary Surgeons in a report describing the reasons for their being conducted and their welfare consequences, and by others.

The term mutilatory generally connotes some form of disfigurement or even maiming. There are multiple definitions and interpretations that carry varying degrees of emotional intensity. Merriam-Webster defines "mutilate" as "to cut up or alter radically so as to make imperfect", but gives a relatively mild example: "the child mutilated the book with his scissors". Animal rights advocates often pejoratively refer to these procedures as mutilations. PETA states that one issue with current forms of non-human animal treatment is that the animals "are mutilated and confined to tiny cages so that we can kill them and eat them."

The Royal College of Veterinary Surgeons noted that the term mutilation is often an emotive one, having implications in common usage of maiming and disfigurement. They stated that there was no satisfactory alternative term that would suffice for their purposes. Their definition is a narrower one: "covering all procedures, carried out with or without instruments which involve interference with the sensitive tissues or the bone structure of an animal, and are carried out for non-therapeutic reasons." The following table contains procedures performed on domesticated animals that may or may not have a purported therapeutic purpose.

Invasive procedures on animals Species Procedures Cats

- Castration
- Claw removal
- Devocalization
- Tail docking
- Tattoo
- Tendonectomy

Cattle

- Castration
- Branding
- Dehorning
- Ear tagging
- Nose ringing
- Restraint
- Tail docking
- Tongue resection (calves)

Dogs

- Castration
- Devocalization/Debarking
- Dewclaw removal

- Ear cropping
- Tail docking
- Tail nicking[a]
- Tattooing
- Teeth removal

Ferrets

- De-scenting[b]
- Tattooing

Horses

- Castration
- Branding
- Pin firing[c]
- Tail blocking[d]
- Tail docking
- Tail nicking[e]

Laboratory mice

- Ear tagging
- Earmark (agriculture)
- Tail clipping[f]
- Tattooing
- Toe clipping[g]

Pigs

- Castration
- Ear docking
- Ear tagging
- Earmarking
- Nose ringing
- Tail docking
- Tattooing

- Teeth clipping
- Tusk trimming

Poultry

- Beak-trimming
- Blinders[h]
- Caponizing
- Desnooding[i]
- Detoeing
- Devoicing
- Dewinging
- Dubbing[j]
- Foie gras force feeding
- Feather plucking
- Pinioning[k]
- Spur removal
- Toe clipping
- Wing clipping

Sheep

- Castration
- Ear tagging
- Earmarking
- Dehorning
- Marking[l]
- Mulesing[m]
- Tail docking
- Teeth grinding

Skunks

- De-scenting

Snake

- Venom duct removal

Prawns/Shrimp

- Eyestalk ablation[n]

Various[o]

- Artificial insemination
- Castration/(caponization)
- Earmarking
- Ear tagging
- Tattooing

VETERINARY DENTISTRY

Veterinary dentistry is the field of dentistry applied to the care of animals. It is the art and science of prevention, diagnosis, and treatment of conditions, diseases, and disorders of the oral cavity, the maxillofacial region, and its associated structures as it relates to animals.

In the United States, veterinary dentistry is one of 20 veterinary specialties recognized by the American Veterinary Medical Association (AVMA). Veterinary dentists offer services in the fields of endodontics, oral and maxillofacial radiology and surgery, oral medicine, orthodontics, pedodontics, periodontics, and prosthodontics. Similar to human dentists, they treat conditions such as jaw fractures, malocclusions, oral cancer, periodontal disease, stomatitis, and other conditions unique to veterinary medicine (e.g. feline odontoclastic resorptive lesions).

Some animals have specialized dental workers, such as equine dental technicians, who conduct routine work on horses.

he practice of veterinary dentistry and oral medicine and surgery is performed by veterinarians in accordance with their state veterinary practice acts. Veterinary health-care workers may be allowed to perform certain non-invasive, non-surgical oral and dental procedures under the direct supervision of a licensed veterinarian in accordance with state regulations. As with other areas of veterinary practice, veterinary dentistry requires a veterinarian-client-patient relationship to protect the health, safety, and welfare of animals.

One of the most vital areas of veterinary dentistry is that it addresses periodontal disease, the most common dental condition in dogs and cats. Pets as young as three years old can show early evidence of periodontal disease, which will worsen if effective preventive measures are not taken. Early detection and treatment are critical, because advanced periodontal disease can cause severe problems and pain.

Pain originating from dental problems is very rarely recognized by owners or professionals. Seldom will an animal become anorexic due to a dental problem. The exception to this is in the case of severe soft tissue injury, for example chronic gingivostomatitis. In general, dental pain is a chronic pain, and it is only after treatment that an owner reports how much better their pet is doing. Pain is often mistaken for a pet just getting old. Very few clients examine their pets' teeth unless they are carrying out daily home care, so actual dental problems often go unnoticed.

Recognizing symptoms that may have a link to dental diseases, such as a nasal discharge or external facial swellings, is often a priority. In some cases, dental patients may present with what appear to be neurological symptoms.

In horses, continuous grazing leads to dental attrition. Also, due to the continuous development of a horse's hypsodont teeth, there will be dental crown wear which can create jagged and sharp edges and attribute to uneven wear. Dental care for horses is designed to prevent "quidding," or dropping food out of the mouth while eating, ulcerations on the cheek or tongue, resistance or unusual sensitivity to the bit, root abscessation, development of mandibular fistulas from infections in the lower cheek teeth, and difficulty flexing at the poll.

ORAL DISEASE

Periodontal disease

Dog with periodontal disease

The most common and significant oral disease is the inflammation of the deeper supporting structures of the tooth and surrounding tissues of the periodontium; it is also called periodontal disease or periodontitis. It begins with the formation of plaque, specifically subgingival plaque within the gingival sulcus or periodontal pocket. This allows a proliferation of bacteria; the subsequent inflammation and the animal's own immune response starts the progression of periodontal disease. The hallmark feature of periodontitis is attachment loss of the tooth from the alveolar bone. Periodontitis is an irreversible process unless the animal is treated with advanced periodontal surgery techniques.

Consequences of periodontal disease

Periodontal disease eventually culminates in tooth loss; however, significant health problems can precede this. Local consequences include the development of an oronasal fistula or periodontal-endodontic lesion, infections or abscesses in the eye, a fractured jaw, osteomyelitis, and oral cancer.

Systemic consequences affect multiple organs. For example, bacteremia from periodontal disease can cause inflammation of the hepatic tissue, portal vein fibrosis, and cholestasis. Chronic stimulation of the immune system can lead to immune complexes in the kidney, causing glomerulonephritis, chronic kidney inflammation and secondary scarring, decreased kidney function and ability to filtrate, and chronic kidney disease. Bacteremia also affects the heart, because circulating bacteria attach to the heart valves, putting the heart at increased risk of endocarditis, hypertension, and roughening of the epithelium of the heart tissue. There may be metabolic changes such as insulin resistance due to the increased inflammatory proteins.

Gingivitis

A canine with gingivitis

Gingivitis is the earliest stage of periodontal disease. It is the inflammation of the gingiva and is caused by bacterial plaque, but it is reversible and preventable.

Gingivitis appears as a thin red line along the margin of the gums and may be accompanied by swollen gum margins, bad breath, plaque and tartar. The earliest clinical sign of gingivitis is bleeding on probing, chewing, or brushing.

Signs and symptoms of oral disease

The most common signs of oral disease include:

- Halitosis
- Broken or discoloured teeth
- Changes in eating behaviour
- Rubbing or pawing at the face
- Ptyalism
- Bleeding from the mouth
- Inability or unwillingness to open or close the mouth
- Change in temperament
- Morbidity
- Weight loss

Diagnosis

Most pet owners are not aware that their pet has an oral problem, so an examination of the oral cavity should form part of every physical examination given by the veterinarian. The current Canine Preventive Healthcare Guidelines and Feline Preventive Healthcare Guidelines from the American Animal Hospital Association (AAHA) and AVMA both include dental care as part of the assessment during annual veterinary examinations. Oral examination in a conscious animal can only give limited information and a definitive oral examination can only be performed under general anesthesia. Examinations are supposed to occur at least once a year to identify any problems and ensure optimal oral health.

Examining the whole animal, even when the primary complaint is the mouth, is routine. Some dental diseases may be the result of a systemic problem and some may result in systemic complications, including kidney, liver, and heart muscle changes. In all cases, dental procedures require a general anesthetic, so it is important to establish the cardiovascular and respiratory status and the physiological values of the patient to avoid risks or complications.

Radiography

Radiographs (x-rays) may be needed to evaluate the health of the jaw and the tooth roots below the gum line. Most dental disease occurs below the gum line and is not visible. Some of the indications for dental radiography include:

- Assessment of bone levels and type of bone loss caused by periodontal disease.
- Assessment of endodontic disease including apical pathology, pulp exposures, and draining fistulae.
- Pathology of the oral soft and hard tissues, including tumors and fractures.
- Evaluation of temporomandibular joint dysfunction.
- Crown/root pathology including tooth resorptive lesions, crown or root fractures, extra roots, or dilacerated roots.
- Comparison of pre/post tooth extraction sites.
- Root canal therapy.
- Oligodontia (absence of most teeth) or extra teeth.
- Assessment of tooth development and chronological dental age of the animal.

Oral abnormalities, anomalies, and defects

Abnormal, anomalous, or defective pet teeth are sometimes encountered in veterinary practice. One or several teeth may be involved and the concern may be for esthetics, function, patient comfort, or a combination of these issues. Causes may be congenital, developmental, or due to lifestyle factors.

Malocclusions

Malocclusion is the imperfect positioning of the teeth when the jaw is closed. In dogs and cats with normal occlusion, the upper incisors rest in front of the lower incisors, the lower canines fit in the diastema between the upper canine and third incisor, the upper first premolars fit behind the lower first premolars, and the upper fourth premolars overlap the lower first molars. Any deviations are known as malocclusions, and they are separated by class.

Class I malocclusion (MAL/1)

Also known as neutrocclusion, MAL/1 occurs when the maxilla and mandible are correctly proportioned, but one or more teeth are misaligned. This type of malocclusion is further classified by type:

- **Rostral cross bite (RXB)** – one or more of the upper incisors are displaced so they rest behind the lower incisors, rather than in front. May be caused by retained deciduous (baby) upper incisors, preventing normal eruption of adult incisors.
- **Caudal cross bite (CXB)** – the mandible is wider than the maxilla in the area of the premolars. Instead of the upper fourth premolar resting against the inner cheek (buccal occlusion) and the lower first molar resting against the tongue (lingual occlusion), the positional relationship is reversed. This is commonly observed in dolichocephalic head types (having a long skull).
- **Linguoversion (LV)** – the lower canines are in the correct anatomic position but are angled inward towards the tongue, which can cause trauma to the palatal tissue (the roof of the mouth).
- **Mesioversion (MV)** – occurs when a tooth is in the correct anatomic position in the dental arch but is angled more forward than normal.
- **Labioversion (LABV)** – occurs when an incisor or canine tooth is in the correct anatomic position in the dental arch but is angled towards the lips.
- **Distoversion (DV)** – occurs when a tooth is in the correct anatomic position in the dental arch but is angled more behind than normal.

- **Buccoversion (BV)** – occurs when a tooth is in the correct anatomic position in the dental arch but is angled outward towards the cheeks.

Class II malocclusion (MAL/2)

Also called distoclusion, brachygnathism, overshot jaw, overbite, and parrot mouth, MAL/2 occurs when the upper teeth rest in front of the lower equivalents. The maxilla is forward (maxillary prognathism) and the mandible is behind (mandibular retrognathism). It is more common in animals with dolichocephalic skulls, such as Collies. It is the most common oral birth defect in horses.

Class III malocclusion (MAL/3)

English bulldog with underbite

Also called mesioclusion, prognathism, undershot jaw, and underbite, MAL/3 occurs when the upper teeth rest behind the lower equivalents. The maxilla is behind (maxillary retrognathism) and the mandible is forward (mandibular prognathism). It is more common in animals with brachycephalic skulls, such as pugs. This type of malocclusion is also often associated with rostral cross bite.

Other malocclusions

- **Level bite** – end-to-end bite of the incisors. Genetically is a degree of prognathism.
- **Wry mouth** – refers to a variety of unilateral occlusal abnormalities. Genetically only affects one quadrant of the mandible or maxilla, such as one segment of the jaw being disproportionately sized relative to the other half.
- **Oligodontia** – only a few teeth are present
- **Anodontia** – congenital absence of teeth
- **Hypodontia** – one or a few teeth are missing
- **Polydontia** – presence of more teeth than is normal (supernumerary teeth)
- **Retained deciduous (baby) teeth** – occurs when erupting permanent teeth do not push deciduous teeth out. This is common in toy breed dogs.

Oral lesions and masses

Some animals will develop oral lesions, masses, or growths, which may be benign or malignant. These may be caused by tooth or gum infections, tumors, or genetic predisposition. Most pets do not show signs of oral masses until it has grown large enough to make chewing and swallowing difficult. Bad breath, excessive drooling, or bloody oral discharge may also be signs of an oral lesion or mass.

MALIGNANT TUMORS

- **Melanoma** – cancerous tumor that spreads to regional lymph nodes and lungs. Bone destruction is usually evident around the tumor. Most commonly observed in dogs; it is rare in cats.
- **Squamous cell carcinoma** – fast-growing tumor that is often ulcerated. It spreads slowly and invades bone tissue. It is the most common oral tumor in cats and second most common tumor in dogs.
- **Fibrosarcoma** – occurs at a younger age than other oral malignancies. It metastasizes slowly but is aggressively invasive. It is the second most common tumor in cats and third most common in dogs.
- **Osteosarcoma** – may affect the bones of the maxilla or mandible. Oral osteosarcoma is more responsive to surgical intervention than the appendicular version.

Nonmalignant tumors

- **Epulis** – general term for any gingival mass. A biopsy and radiograph is required in order to differentiate the type.
- **Peripheral odontogenic fibroma** – grows from the periodontal ligament. It contains no invasion of bone tissue and has a firm, smooth surface. It may cause tooth displacement. It is the most common benign tumor in dogs.
- **Peripheral acanthomatous ameloblastoma** – a slow-growing tumor that has a raised, cauliflower appearance and may transform into a malignant tumor at a later stage. It is usually invaded by surrounding bone tissue and aggressive surgical removal is required.

Resorptive lesions

Also known as feline cervical neck lesions or feline odontoclastic resorptive lesions, resorptive lesions are most commonly observed in the cat but have been identified in other species as well. The lesion usually starts at the cementoenamel junction and furcation area between the roots of a tooth. They appear as an overgrowth of gingival or pulpal tissue. The lesions erode the dentin within a single tooth (or several simultaneously). It spreads rapidly once it reaches the pulp of the tooth. The crown of the tooth may appear normal, but the tooth may have little to no root left. This is a very painful condition.

Teeth with resorptive lesions are incurable and cannot be prevented. There is no concrete evidence for the cause, but some studies have suggested they are caused by an excess of vitamin D. Extraction is the only method of treatment.

Developmental conditions

- **Gemination** – two crowns share one single root canal in an attempt to form two separate teeth from one enamel origin.
- **Fusion** – two tooth buds grow together to form one larger tooth.
- **Impaction** – the inability of the tooth to erupt through the gum. This can cause the development of a fluid-filled cyst surrounding the tooth and destruction to alveolar bone.
- **Misdirected teeth** – teeth that erupt in an abnormal direction.
- **Enamel dysplasia** – insufficient hardness or amount of enamel due to not properly forming. The enamel is soft, flakes off, stains easily, has a rough or pitted surface, and exposes underlying dentin once chipped away.

Dental cleaning

In dogs, as in humans, daily tooth brushing is considered the gold standard for at-home prophylaxis and prevention of gingivitis and periodontal disease progression. A Swedish study with over 60,000 respondents reported that only 4% of dog owners brushed their dog's teeth daily.

Basic dental cleaning under general anesthesia includes scaling to remove dental plaque, tartar, and calculus deposits, as well as polishing to smooth out microabrasions caused by the dental equipment and normal wear and tear. Endodontic procedures such as tooth extractions and root canals are also performed.

The horse will be sedated or given analgesics for dental cleaning. Dental prophylaxis is often called "floating," and handheld rasps or power instruments are used to grind, balance, and realign the occlusal surfaces of the incisors and cheek teeth. Motorized dental instruments should be used carefully to avoid thermal and pressure trauma to the teeth. This means using low-speed grinders without leaving them in one spot for long periods of time, using light pressure, and using water irrigation to rinse and prevent overheating caused by the friction of the instruments.

DENTAL INSTRUMENTS

Dental instruments are tools used to provide dental treatment. They include tools to examine, manipulate, treat, restore, and remove teeth and surrounding oral structures.

Hand instruments

- **Curette** – used for removing calculus above and below the gumline. It is also used for root planing. It has a U-shaped cross-section and the tip is rounded with up to two sharp edges. The two most commonly used types are the Gracey curette and the Universal curette.
- **Sickle scaler** – used for removing calculus above the gumline, as well as removing calculus from pits, fissures, developmental grooves, and areas between each tooth. The blade can be straight or curved with a sharp tip and two cutting edges on either side.
- **Explorer** – used to examine the surface of the tooth to detect any abnormalities, such as resorptive lesions, caries, or fractured teeth. It is also used to assess tooth mobility. The most common type is the Shepherd's hook.
- **Periodontal probe** – used to measure gingival recession, which ascertains the stage of any periodontal disease. It has a blunted tip that is marked in 1 mm increments. The probe is inserted into the gingival sulcus to measure its depth. Normal sulcus depth in the dog is < 3 mm and < 1 mm in cats.
- **Elevator** – used to stretch, cut, and tear the periodontal ligament in order to displace the tooth root from the socket. The tip has a rounded scoop appearance with a sharp edge, which may or may not be serrated.
- **Luxator** – used to cut the periodontal ligament around the tooth; it is not used for leverage. It is similar to a dental elevator in design but has a thinner tip to allow easier access to the periodontal ligament.
- **Extraction forceps** – used for gripping and removing the tooth after it has been loosened. It can also be used for cracking and dislodging heavy dental calculus.

Power instruments

- **Ultrasonic scaler** – used for removing calculus above the gumline. It utilizes a removable tip that vibrates at high frequencies by converting sound waves into mechanical vibrations, allowing for more rapid removal of calculus. It generates a substantial amount of heat and may cause thermal damage to the pulp if left on the surface of the tooth for too long.

- **Sonic scaler** – used for removing calculus above the gumline. It utilizes compressed air and operates at lower frequencies than an ultrasonic scaler. It is also less likely to cause thermal damage.
- **Low-speed handpiece** – utilizes a rubber tip called a prophy angle that polishes microabrasions and small grooves on the surface of the tooth.
- **High-speed handpiece** – utilizes a variety of removable dental burs to section teeth and remove alveolar bone during extractions.

Dental charting

All findings from an oral examination should be recorded on a dental chart. These include missing, rotated, and fractured teeth, probing depths of gingival recession, enamel hyperplasia or other enamel defects, mobility, furcation involvement, and other oral pathology. Charting not only records the current state of the dentition and soft tissues of the oral cavity, allowing the formulation of a treatment plan, but also provides a permanent record for future comparisons.

The severity of gingivitis is scored by using the Gingival Index (GI), which consists of four stages:

- Stage 0 (GI0) – characterized by normal gingiva.
- Stage 1 (GI1) – marginal gingivitis with mild swelling of the gum tissue, some color change, and no bleeding upon probing.
- Stage 2 (GI2) – moderate swelling and inflammation of the gum tissue, and bleeding upon probing.
- Stage 3 (GI3) – severe gingivitis with marked swelling and inflammation, as well as spontaneous bleeding.

The severity of periodontal disease is scored by using the Periodontal Disease Index (PD), which consists of five stages:

- Stage 0 (PD0) – characterized by the absence of disease.
- Stage 1 (PD1) – characterized by presence of gingivitis. This stage can be treated with dental scaling, polishing, irrigation, and home health care.
- Stage 2 (PD2) – early periodontal disease with less than 25% attachment loss of the tooth from the alveolar bone. The treatment for this stage includes the addition of locally applied antimicrobials and subgingival scaling.
- Stage 3 (PD3) – established periodontal disease with 25–50% attachment loss of the tooth from the alveolar bone. Treatment includes teeth extraction,

closed or open root planing, or advanced periodontal treatment options like guided tissue regeneration.

- Stage 4 (PD4) – advanced periodontal disease with more than 50% attachment loss of the tooth from the alveolar bone. Extraction of teeth or periodontal surgery, including osseous resective or additive procedures, are necessary. The prognosis for this stage is guarded.

The amount of calculus present is scored by using the Calculus Index (CI), which consists of four stages:

- Stage 0 (CI0) – no calculus is present.
- Stage 1 (CI1) – some calculus above the gum line that covers less than 1/3 of the tooth surface.
- Stage 2 (CI2) – moderate calculus covering 1/3 to 2/3 of the tooth surface with minimal calculus below the gum line.
- Stage 3 (CI3) – heavy calculus covering more than 2/3 of the tooth surface and extending below the gum line.

Equine dentistry

Equine dentistry is the practice of dentistry in horses, involving the study, diagnosis, prevention, and treatment of diseases, disorders and conditions of the oral cavity, maxillofacial area and the adjacent and associated structures.

The practice of equine dentistry varies widely by jurisdiction, with procedures being performed by veterinary physicians (both in general and specialist practice), specialist professionals termed equine dental technicians or equine dentists, and by amateurs, such as horse owners, with varying levels of training.

In some jurisdictions, the practice of equine dentistry, or specific elements of equine dentistry, may be restricted only to specialists with specified qualifications or experience, whereas in others it is not controlled.

History

Equine dentistry was practiced as long ago as 600 BCE in China, and has long been important as a method of assessing the age of a horse. This was also practiced in ancient Greece, with many scholars making notes about equine dentistry, including Aristotle with an account of periodontal disease in horses in his History of Animals, and in Rome with Vegetius writing about equine dentistry in his manuscript "The Veterinary Art".

In later years, the importance of dentition in assessing the age of horses led to veterinary dentistry being used a method of fraud, with owners and traders altering the teeth of horses to mimic the tooth shapes and characteristics of horses younger than the actual age of the equine.

The first veterinary dental school was founded in Lyon, France, in 1762 and created additional knowledge about dentistry in horses and other animals.

Equine dental technicians

Equine dental technicians (also known colloquially as equine dentists, although this is not reflective of their official title) are veterinary paraprofessionals who specialize in routine dental work on horses, especially procedures such as rasping the sharp edges of teeth, also known as 'floating'.

Scope of practice may be dictated by statute. For instance, in the United Kingdom, any person, without any qualification may examine and rasp horses' healthy teeth with manual tools, remove deciduous caps (baby teeth) or remove supragingival calculus, whereas only qualified equine dental technicians or vets may remove teeth, rasp fractured teeth and use motorized dental instruments.

Relationship between vets and lay practitioners

There has been a long history of animosity between fully qualified vets and those lay people performing equine dentistry. This has led in some cases to an increase in voluntary and regulatory schemes in the sector. In the UK in the early 1990s, the veterinary profession engaged with lay practitioners to establish a formal system of examination and a register of qualified professionals. Whilst not compulsory, qualified persons were given an extended scope of practice for procedures normally reserved to vets.

In the United States, lay practitioners have been subject to legal action to gain injunctions against their practice, with case law in states including Missouri effectively precluding the practice of equine dentistry by persons other than vets.

Horse teeth

Horse teeth refers to the dentition of equine species, including horses and donkeys. Equines are both heterodontous and diphyodontous, which means that they have teeth in more than one shape (there are up to five shapes of tooth in a horse's mouth), and have two successive sets of teeth, the deciduous ("baby teeth") and permanent sets.

As grazing animals, good dentition is essential to survival. Continued grazing creates specific patterns of wear, which can be used along with patterns of eruption to estimate the age of the horse.

TYPES OF TEETH

A fully developed horse of around five years of age will have between 36 and 44 teeth. All equines are heterodontous, which means that they have different shaped teeth for different purposes. All horses have twelve incisors at the front of the mouth, used primarily for cutting food, most often grass, whilst grazing. They are also used as part of a horse's attack or defence against predators, or as part of establishing social hierarchy within the herd.

Immediately behind the front incisors is the interdental space, where no teeth grow from the gums. This is where the bit is placed, if used, when horses are ridden.

Behind the interdental space, all horses also have twelve premolars and twelve molars, also known as cheek teeth or jaw teeth. These teeth chew food bitten off by incisors, prior to swallowing.

In addition to the incisors, premolars and molars, some, but not all, horses may also have canine teeth and wolf teeth. A horse can have between zero and four canine teeth, also known as tusks (tushes for the deciduous precursor), with a clear prevalence towards male horses (stallions and geldings) who normally have a full set of four. Fewer than 28% of female horses (mares) have any canine teeth. Those that do normally only have one or two, and these may be only partially erupted.

Between 13 and 32% of horses, split equally between male and female, also have wolf teeth, which are not related to canine teeth, but are vestigial first premolars. Wolf teeth are more common on the upper jaw, and can present a problem for horses in work, as they can interfere with the bit. They may also make it difficult during equine dentistry work to rasp the second premolar, and are frequently removed.

Tooth eruption

Horses are diphyodontous, erupting a set of first deciduous teeth (also known as milk, temporary, or baby teeth) soon after birth, with these being replaced by permanent teeth by the age of approximately five years old. The horse will normally have 24 deciduous teeth, emerging in pairs, and eventually pushed out by the permanent teeth, which normally number between 36 and 40. As the deciduous teeth are pushed up, they are termed "caps". Caps will eventually shed on their own, but may cause discomfort when still loose, requiring extraction.

It is possible to estimate the age of a young horse by observing the pattern of teeth in the mouth, based on which teeth have erupted, although the difference between breeds and individuals make precise dating impossible.

Common ages for tooth eruption.

Type of tooth	Number	Deciduous	Permanent
Incisor	First (central)	birth to 8 days	2.5 yrs
Incisor	Second (intermediate)	4.5–6 weeks	3.5–4 yrs
Incisor	Third (corner)	6–9 months	4.5–5 yrs
Canine		Absent	3.5–5 yrs, some around 6 yrs (if ever)
Premolar	First (wolf)	Absent	6 months to 3 years (if ever)
Premolar	Second	birth to 2 weeks	2–3 yrs
Premolar	Third	birth to 2 weeks	2.5–3 yrs
Premolar	Fourth	birth to 2 weeks	3–4 yrs
Molar	First	Absent	9–12 months
Molar	Second	Absent	2 yrs
Molar	Third	Absent	3–4 yrs

All teeth are normally erupted by the age of five, at which point the horse is said to have a "full mouth", but the actual age this occurs will depend on the individual horse, and also by breed, with certain breeds having different average eruption times. For instance, in Shetland ponies the middle and corner incisor tend to erupt late, and in both draft horses and miniature horses, the permanent middle and corner incisors are usually late appearing.

- Horse dentition over time

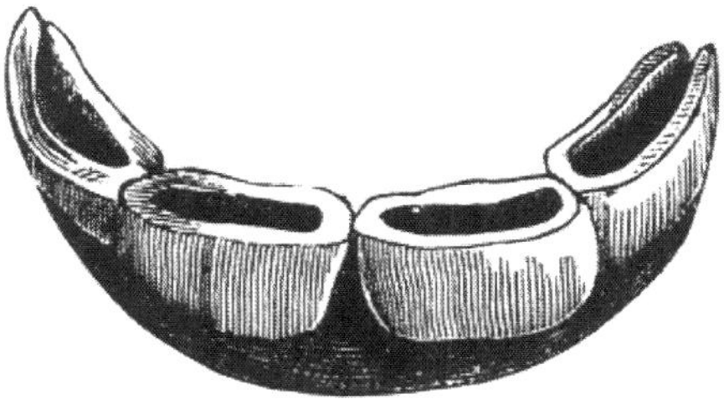

Fig. 3.18 Foal

Fig. 3.19 Yearling

Fig. 3.20 Two years

Fig. 3.21 Three years

Fig. 3.22 Five years

Fig. 3.23 Six years

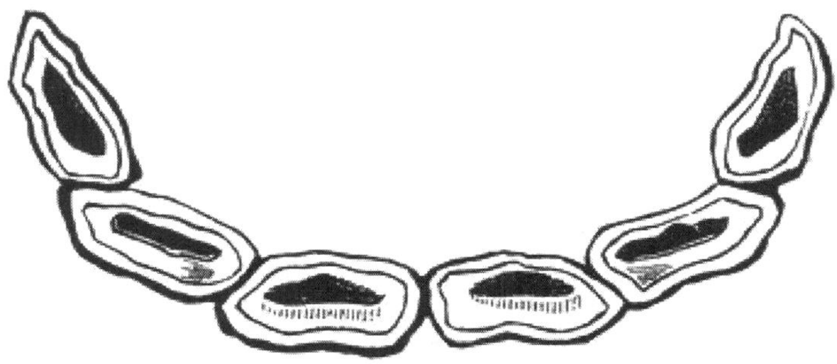

Fig. 3.24 Seven years

Fig. 3.25 Nine years

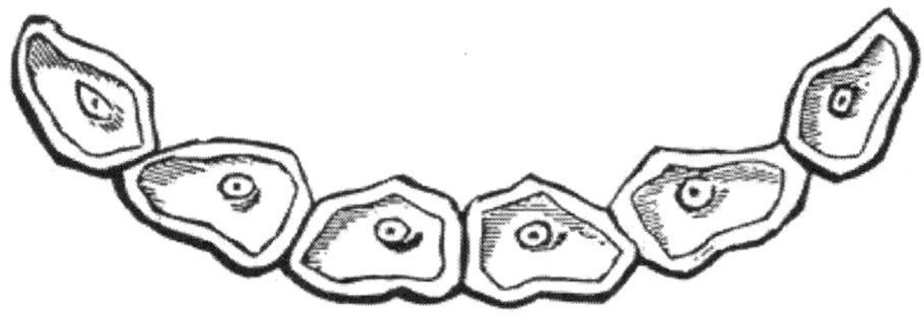

Fig. 3.26 Twelve years

Tooth wear

The incisors, showing the wear and marks on their tables.

Horse teeth often wear in specific patterns, based on the way the horse eats its food, and these patterns are often used to conjecture on the age of the horse after it has developed a full mouth. As with aging through observing tooth eruption, this can be imprecise, and may be affected by diet, natural abnormalities, and vices such as cribbing.

The importance of dentition in assessing the age of horses led to veterinary dentistry techniques being used as a method of fraud, with owners and traders altering the teeth of horses to mimic the tooth shapes and characteristics of horses younger than the actual age of the equine.

Equine teeth have evolved to wear against the tooth above or below as the horse chews, thus preventing excess growth. The upper jaw is wider than the lower one. In some cases, sharp edges can occur on the outside of the upper molars and the inside of the lower molars, as they are unopposed by an opposite grinding surface. These sharp edges can reduce chewing efficiency of the teeth, interfere with jaw motion, and in extreme cases can cut the tongue or cheek, making eating and riding painful.

In the wild, natural foodstuffs may have allowed teeth to wear more evenly. Because many modern horses often graze on lusher, softer forage than their ancestors, and are also frequently fed grain or other concentrated feed, it is possible some natural wear may be reduced in the domestic horse. On the other hand, this same uneven wear in the wild may have at times contributed to a shorter lifespan. Modern wild horses live an estimated 20 years at most, while a domesticated horse, depending on breed and management, quite often lives 25 to 30 years. Thus, because domesticated animals also live longer, they may simply have more time to develop dental issues that their wild forebears never faced.

TYPICAL WEAR PATTERNS

Typical Wear Patterns in Horses Include

- **Cups:** are hollow and rectangular or oval in shape, appearing on the tables of the permanent incisors, that wear away over time. In general, cups are worn away on the lower central incisors by age 6, the lower intermediates by age 7, and corners at age 8. The cups of the upper central incisors are worn away by 9 years of age, the upper intermediate incisors by 10, and the corners by 11. When all the cups are gone, the horse is referred to as smooth mouthed. In the past, dishonest dealers would "bishop" the teeth of old horses, usually by burning an indentation into the teeth, to imitate cups: but this practice was detectable by the absence of the white edge of enamel that always surrounds the real cup, by the shape of the teeth, and other marks of age about the animal.
- **Pulp mark/Dental star:** After some wear has occurred on the teeth, the central pulp cavity is exposed, and the tooth is marked by a "dental star" or "pulp mark" that is smaller than the incisor cups. These begin as a dark line in front of the dental cup, which grows in size and becomes more oval in shape as the cups are worn away. Dental stars are usually first visible at age 6, on the animal's lower central incisors, and very visible by age 8. They appear on the lower intermediates by age 9, and on the other incisors between the ages of 10–12 years.
- **Hook/Notch:** A hook appears on the upper corner incisor around age 7, and disappears by age 8. It reappears around age 13, again disappearing about 1 year later.
- **Galvayne's Groove:** The Galvayne's groove occurs on the upper corner incisor, producing a vertical line, and is helpful in approximating the age of older horses. It generally first appears at age 10, reaches halfway down the tooth by age 15, and is completely down the tooth at age 20. It then begins to disappear, usually half-way gone by age 25, and completely gone by age 30.
- **Angle and Shape of the incisors:** As the horse ages, the angle of the incisors generally becomes more acute, slanting forward. The incisors gradually change their form as the horse ages, becoming round, oval, and then triangular.

CONTINUOUS ERUPTION AND LOSS

A horse's incisors, premolars, and molars, once fully developed, continue to erupt as the grinding surface is worn down through chewing. A young adult horse's teeth are typically 4.5–5 inches long, but the majority of the crown remaining below the gumline in the dental socket. The rest of the tooth slowly emerges from the jaw, erupting about 1/8" each year, as the horse ages. When the animal reaches old age, the crowns of the teeth are very short and the teeth are often lost altogether. Very old horses, if lacking molars to chew, may need soft feeds to maintain adequate levels of nutrition.

Older horses may appear to have a lean, shallow lower jaw, as the roots of the teeth have begun to disappear. Younger horses may seem to have a lumpy jaw, due to the presence of permanent teeth within the jaw.

The teeth and the bit

Simple snaffle bit fitted to a horse, behind the incisors, but in front of the premolars

If a bit is fitted to a horse, along with a bridle, the normally metal bar of the bit lies in the interdental space between the incisors (or canines, where present) and premolars. If the bridle is adjusted so that the bit rests too low, or too high, it may push against the teeth and cause discomfort.

Sometimes, a "bit seat" is filed in the first premolar, where the surface is rounded so that the flesh of the cheek is not pushed into the sharp edge of the tooth, making riding more comfortable for the horse, although the practice is controversial.

Dental problems

Modern equine dentistry. This horse is heavily sedated and has been given analgesics, its head is supported by a sling. The mouth is kept open with a (horse) mouth gag, commonly referred to as a "speculum".

Like all mammals, horses can develop a variety of dental problems, with a variety of dental services available to minimise problems through reactive or prophylactic intervention.

Equine dentistry can be undertaken by a vet or by a trained specialist such as an equine dental technician, or in some cases is performed by lay persons, including owners or trainers.

Problems with dentition for horses in work can result in poor performance or behavioural issues, and must be ruled out as the cause of negative behaviours in horse. Most authorities recommend regular checks by a professional, normally six monthly or annually.

Problems due to wear patterns

A wolf tooth, located just in front of the premolars.

The wear of the teeth can cause problems if it is uneven, with sharp points appearing, especially on the outer edge of the molars, the inner edge of the premolars and the posterior end of the last molars on the bottom jaw.

Other specific conditions relating to wear include a "step mouth", where one molar or premolar grows longer than the others in that jaw, normally because the corresponding tooth in the opposite jaw is missing or broken, and therefore could not wear down its opposite, a "wave mouth", where at least two molars or premolars are higher than the others, so that, when viewed from the side, the grinding surfaces produce a wave-like pattern rather than a straight line, leading to periodontal disease and excessive wear of some of the teeth, and a "shear mouth" when the grinding surfaces of the molars or premolars are severely sloped on each individual tooth (so the inner side of the teeth are much higher or lower than the outer side of the teeth), severely affecting chewing.

Horses may also experience an overbite/brachygnathism (parrot mouth), or an underbite/prognathism (sow mouth, monkey mouth). These may affect how the incisors wear. In severe cases, the horse's ability to graze may be affected. Horses also sometimes suffer from equine malocclusion where there is a misalignment between their upper and lower jaws.

The curvature of the incisors may also vary from the normal, straight bite. The curvature may be dorsal or ventral . These curvatures may be the result of an incisor malocclusion (e.g. ventral=overbite/dorsal=underbite). The curvature may also be diagonal, stemming from a wear pattern, offset incisors, or pain in the cheek teeth (rather than the incisors), which causes the horse to chew in one direction over the other.

Other dental problems

Other common problems include abscessed, loose, infected, or cracked teeth, retained deciduous teeth, and plaque buildup. Wolf teeth may also cause problems, and are many times removed, as are retained caps.

Prevention of dental problems

Many veterinarians will use a Triadan chart to record the animal's dental problems for future reference.

To help prevent dental problems, it is recommended to get a horse's teeth checked by a vet or equine dental technician every 6 months. However, regular checks may be needed more often for individuals, especially if the horse is very young or very old. Additionally, the horse's teeth should be checked if it is having major performance problems or showing any of the above signs of a dental problem.

Many horses require floating (or rasping) of teeth once every 12 months, although this, too, is variable and dependent on the individual horse. The first four or five years of a horse's life are when the most growth-related changes occur and hence frequent checkups may prevent problems from developing. Equine teeth get harder as the horse gets older and may not have rapid changes during the prime adult years of life, but as horses become aged, particularly from the late teens on, additional changes in incisor angle and other molar growth patterns often necessitate frequent care. Once a horse is in its late 20s or early 30s, molar loss becomes a concern. Floating involves a veterinarian wearing down the surface of the teeth, usually to remove sharp points or to balance out the mouth. However, the veterinarian must be careful not to take off too much of the surface, or there will not be enough roughened area on the tooth to allow it to properly tear apart food. Additionally, too much work on a tooth can cause thermal damage (which could lead to having to extract the tooth), or expose the sensitive interior of the tooth (pulp). A person without a veterinary degree who performs this service is called a horse floater or equine dental technician.

In popular culture

The common folk saying "don't look a gift horse in the mouth" is taken from the era when gifting horses was common. The teeth of a horse are a good indication of the age of the animal, and it was considered rude to inspect the teeth of a gifted animal as you would one that you were purchasing. The saying is used in reference to being an ungrateful gift receiver.

DENTAL HEALTH DIETS FOR DOGS

Oral disease is one of the most common diseases found in dogs. It is caused by the buildup of various anaerobic bacteria in the mouth which forms plaque, eventually hardening into tartar on the teeth along the gum line, and is related to the development of gingivitis. Since small and toy breeds have a much smaller jaw but contain the same number of teeth, crowding allows higher bacterial build up and puts them at higher risk of developing periodontal disease.

Although daily brushing provides the best preventative measure, feeding a dental diet or using dental chews for dogs is an effective approach pet owners can take to help prevent and control accumulation of plaque and tartar to avoid consequences of severe periodontal disease. Dental diets consider many aspects of oral health such as inflammation, bone health and physical and chemical characteristics of the kibble when being formulated. Some diets, chews and/or other oral products may also strive to receive approval by the Veterinary Oral Health Council (VOHC) for validation of their product's claims.

Bone health

Teeth play a vital role in an animal's survival; they are used for eating, grooming and defense.

Each tooth is made up of a crown, which is above the gum line and covered in enamel, and roots that anchor the tooth to the alveolar bone. Beneath the enamel, there are collagen fibres and inorganic hydroxyapatite, which together form dentin. Hydroxyapatite is the storage form of calcium found in the bone.

Blood calcium concentration are kept within a narrow range. If deficient, calcium will be mobilized from the bone to compensate imbalances.

The ratio of calcium to phosphorus is essential to bone and teeth health; if there is excessive phosphorus within a diet, it can result in resorption and reduction of the alveolar bone as shown in Henrikson's studies when dogs were fed a low calcium, high phosphorus diet.

Vitamin D also plays a major role in calcium absorption and bone health because it helps to maintain calcium homeostasis and bone metabolism. Deficiencies negatively impact parts of the teeth including the gingivae, periodontal ligaments and the alveolar bone, leading to degeneration of the teeth and jaw.

According to the Association of American Feed Control Officials (AAFCO) dietary recommendations based on dry matter content, the maximum amount of calcium within a diet is 1.8% and the phosphorus content within a diet cannot exceed 1.6%. These recommendations ensure that the calcium to phosphorus ratio will be 1:1, or at max 2:1. AAFCO also states that the maximum amount of vitamin D within a canine diet cannot exceed 3000 IU/kg.

Inflammation

One of the main aims of dental food for dogs is to minimize plaque accumulation and gingival inflammation. The oral cavity of dogs can be exposed to a variety of bacteria, parasites and viruses from their environment and it is not surprising that these induce an inflammatory reaction in the periodontium. The accumulation of these bacteria, along with extracellular polysaccharides and salivary glycoproteins, leads to the formation of plaque. This bacteria buildup in plaque causes gingival inflammation and indirectly stimulates activation of the host immune system. Severe gingivitis in dogs can further advance into periodontal disease in which the periodontal tissues begin to degrade and, if left untreated, can lead to tooth loss. Studies have also shown that periodontal disease can negatively affect systemic health which impacts the overall health of the dog, therefore exhibiting the importance of preventing periodontal disease. Using a dental diet may be beneficial towards preventative care in regards to periodontal disease. Dental diets not only try to combat the buildup of plaque physically, but attempt to reduce plaque build up chemically as well. Chemical compounds such as anti-microbial elements are added to some dental diets to directly target the oral cavity in order to prevent gingivitis.

Chemical compounds in kibble

Excessive plaque formation can lead to severe inflammation of the gums which may result in the loss of teeth. This plaque is also able to mineralize on the teeth, creating a crystallized calcium phosphate build up known as tartar. Ingredients such as polyphosphate, zinc salts, and chlorhexidine have active properties to help prevent or reduce the formation of plaque and tartar. These compounds are able to work on the whole mouth including between teeth and all the way up to the gum line, and is not limited to just the chewing contact surface area.

Polyphosphates constitute a wide variety of compounds that have beneficial effects on oral health, the most effective of these are the hexametaphosphates, tripolyphosphates and pyrophosphates and may be attached with an alkali metal, alkaline metal, ammonium or a salt. Sodium tripolyphosphates as an example, are able to bind directly to the enamel as well as the mineralized sections of the teeth where it takes up the binding sites for calcium to attach, and interacts with calcium already bonded to enamel, to prevent it being able to form crystals. Calcium mixed in the saliva can also be bound by sodium tripolyphosphates thus further preventing crystal formation. Sodium tripolyphosphates are also beneficial for stain prevention, and can be broken down in the digestive tract and provide a source of phosphorus and allows the undeposited calcium to be available for whole body metabolism.

Zinc compounds, more specifically zinc ascorbate, also play a role in preventing plaque accumulation due to antimicrobial activity. Zinc salts inhibit bacterial growth by binding to sulfur to control plaque formation, as well as reduce foul oral odours. However, research has only been performed on cats, so the same evidence may not be directly applicable to dogs.

Chlorhexidine is another example of a non-specific antimicrobial agent. It acts by disrupting bacterial deposition and plaque formation on the teeth, further assisting in reducing gingivitis. Chlorhexidine can also reduce the amount of pre-existing plaque. Chlorhexidine is generally found on products such as dental chews or oral rinses and may not be suitable for use in a long-term diet as prolonged use may lead to staining of the teeth and tongue, and change the taste of the food.

Such components of the diet may help prevent plaque accumulation, thus reducing the overall inflammatory state of periodontal disease in dogs.

PHYSICAL CHARACTERISTICS OF KIBBLE

The physical characteristics of dental kibble are very important as they contribute to the mechanical cleaning of the tooth's surface. The kibble need to be large and very dense to promote chewing, as more time spent chewing will aid in the decrease of accumulation of plaque, tartar and calculus. When the kibble is being chewed, the broken pieces rub against the tooth's surface and scrape off the buildup of bacteria that forms the tartar, plaque and calculus. Ingredients that have an adequate amount of carbohydrates and are high in insoluble fibers (typically around 10%), such as soy fiber, wheat bran fiber or cellulose fiber, are added to increase the density of the kibble to assure it does not easily crumble and promotes a texture favourable to scraping of the tooth. Increasing the diameter will also significantly reduce the accumulation of plaque and calculus compared to traditional dog food as it will require more chewing by the molars. Traditional hard kibble will not have the same effects on oral health as it has been proven that the increase in fiber and diameter is what provides those benefits in the diet. A disadvantage is that the mechanical cleansing is more beneficial for the molars compared to the incisors and canines. The incisors and canines are typically used for grabbing and tearing, whereas the molars are used for crushing and chewing, so most of the contact with the broken kibble is in the back of the mouth. The physical characteristics and chemical agents of the kibble work together to delay the buildup of plaque, and are both equally important to the cleansing of the dog's teeth.

Dental health diets for cats

Oral health can be difficult for pet owners and veterinary teams to manage in cats, particularly for pets whose owners are not committed to regular tooth brushing and/or dental treats. Oral disease is common among cats, and may lead to other health issues such as bacterial infections of major organs including the heart, kidneys and liver. When pet owners are aware of the benefits of supporting good oral health in cats, this substantially improves positive outcomes. Dietary selection, along with at-home-dental-hygiene care, allows cat owners to influence the oral status of their pets.

Feline dental health diets are pet foods designed to prevent oral diseases in cats. Many commercial and prescription pet food manufacturers offer dental specific diets aimed to improve the oral health of adult domestic cats.

Rabbit Dental Disease

- Rabbits, guinea pigs, chinchillas, and prairie dogs are herbivorous species.
- They have a very specialized nutrition and nutritional requirements.
- Complications due to dental disease are wider and more severe when compared to those of carnivores.
- The "syndrome" that most rabbit/rodent dental diseases are referred to is called acquired dental disease (ADD).

Clinical exam: Symptoms of dental disease can be varied.

- Poor coat quality
- Weakness
- Painful posture
- Anorexia/dysphagia
- Dehydration
- Reduced or abnormal feces
- Soft or watery feces
- Epiphora
- Dacryocystitis
- Nasal discharge
- Exophthalmos
- Panophthalmitis
- Dewlap, facial, and inguinal dermatitis
- Mandibular and/or maxillary masses

DENTITION OF RABBITS

Dental formula: 2I 0C 3P 3M/1I 0C 2P 3M

Incisors

4 upper incisors: 2 rows of 2 incisors per row; the caudal most upper incisors are much smaller and are called the "peg teeth".

2 lower incisors: In a resting position, the lower incisors occlude with the peg teeth or in a small notch between the rostralmost upper incisors and the peg teeth.

Premolars and Molars

- Collectively referred to as the "cheek teeth"
- No functional difference between premolars and molars
- Each upper arcade has 6 cheek teeth.
- Each lower arcade has 5 cheek teeth.

Roots of the upper cheek teeth lie extremely close to the orbit and the nasolacrimal duct.

- All teeth in the rabbit are elodont (open-rooted). They continuously grow throughout the life of the rabbit.
- In the rostrocaudal view of the skull, the upper and lower cheek teeth are out of occlusion; the upper cheek teeth lie in a more lateral position and the lower cheek teeth are medial.

ACQUIRED DENTAL DISEASE OF THE INCISOR TEETH

Commonly, the lower incisors will elongate and protrude rostrally, while the upper incisors will elongate and curl inwards.

Incisor teeth can appear out of occlusion, at abnormal angles, appearing in various states of deterioration.

Causes

- Congenital (not considered an acquired dental disease, since it is congenital)
- Mandibular/maxillary abnormalities (e.g., tumors, jaw fractures, etc., that disrupt incisor occlusion or growth)
- Secondary to ADD of the cheek teeth
- Traumatic (fractures of the incisor teeth)
- Metabolic (e.g., poor nutrition, lack of sunlight)

ACQUIRED DENTAL DISEASE OF THE CHEEK TEETH

Pathophysiology

- Three main etiologies have been proposed for the pathophysiology of ADD:
- Congenital defects
- Mandibular prognathism
- Maxillary brachygnathism
- Nutrition
- Insufficient wearing of occlusal planes
- Fractures due to inappropriate food
- The diet of pet rabbits is not identical to that of free ranging rabbits
- Even rabbits allowed to graze freely likely do not consume grasses with the same types and amounts of silicates

Nearly 100% of pet rabbits will develop ADD during their lifetime, and many cases will be mild or inapparent

Nearly 100% of the owners feed rabbits improperly

Insufficient wearing leads to excessive elongation of crowns and roots, with bending of the axis of the cheek teeth, deformation of apexes and development of spurs

- Physiology of chewing: 4 basic movements in the rabbit
- Rostrocaudal, caudorostral, lateral, and vertical

The degree of lateral motion of the jaw is partially determined by the type of food consumed

Hay or natural fibrous vegetable food produces wide lateral movements with reduced vertical motion

Concentrated pelleted food or grains produces reduced lateral movements and more extensive vertical motion (more pressure on roots, less wearing)

Metabolic

Metabolic bone disease, a.k.a. nutritional secondary

Hyperparathyroidism (NSHP)

Frances Harcourt-Brown proposed this pathophysiologic hypothesis based on the following observations

Many rabbits who live outdoors with exposure to natural sunlight do not develop ADD

Most rabbits with metabolic bone disease demonstrate demineralization of the skull bones

A study performed on more than 80 rabbits with ADD and MBD demonstrated parathyroid (PTH) level much higher than normal levels

History

The rabbit is prone to hide or minimize clinical symptoms. History is always incomplete, absent, or misleading.

The rabbit is eating nothing or less than normal? How much less than normal?

Complete review of diet offered and consumed

- Quantity/quality of stool
- Does the rabbit chew frequently without food present?
- Is the rabbit interested in food, but dysphagic?
- Is the rabbit drinking?

Physical Exam

Clinical Exam - Without Anesthesia

Thorough dental evaluation must always be performed in rabbits and rodents, even in the absence of significant history supporting dental disease.

It is mandatory to perform inspection of oral cavity always after the clinical exam without sedation.

Clinical Exam - With Anesthesia

In cases of suspected or confirmed acquired dental disease, it is mandatory to perform a complete clinical examination under anesthesia.

During clinical examination under anesthesia, additional diagnostic testing is very important.

BASIC PRINCIPLES OF ANESTHESIA AND ANALGESIA

- Most dental procedures depend on the anesthetic protocol for induction.
- Compare and contrast induction with injectable vs. inhalant drugs.
- Induction with Injectable Anesthesia
- Option 1 (deep sedation)
- Ketamine/butorphanol/dexmedetomidine
- Option 2 (mild-moderate sedation)
- Midazolam/butorphanol
- Other injectable sedation options exist.
- After 5–10 minutes, rabbit should be sufficiently sedated

From there, administer:

- Oxygen (when possible)
- Oxygen + isoflurane at 1%–3% if a greater depth of anesthesia is needed
- Advantages/Disadvantages
- Quicker induction time (2–5 minutes)

Less stressful for most patients

Quicker achievement of surgical anesthetic plane

Respiratory depth and frequency, cardiac activity, blood pressure: more depressed than inhalant agents alone

Longer recovery time (although administering atipamezole at the end of the procedure can be used to reverse the dexmedetomidine and allow for a faster recovery)

- Induction with Inhalant Anesthesia
- Oxygen + isoflurane at 3%–5% (by face mask)
- Advantages/Disadvantages
- Longer induction time
- Constant restraint of the patient needed during the induction phase
- More stressful
- Risk of apnea in rabbits

- After induction, the proper surgical anesthetic plane is still not achieved
- Respiratory depth and frequency, cardiac activity, blood pressure: less depressed than injectable agents
- Quicker recovery time

PATIENT INTUBATION FOR ANESTHESIA - IS IT ALWAYS THE BEST CHOICE?

- Intubation provides more direct delivery of anesthesia to the lungs, provides emergency respiratory access, and can prevent accidental aspiration.
- Rabbits can usually be intubated, with some exceptions; special techniques are required for intubation.
- The endotracheal tube may interfere with taking radiographs of the skull, and with certain dental procedures.
- Nasal intubation, nasotracheal intubation, and orotracheal are all possible options depending on the procedure performed.
- Author's preferred anesthetic regimen for diagnosis, initial evaluation, and treatment of dental disease:

Induction of anesthesia with injectable agents

Maintenance of anesthesia with supplemental isoflurane as needed

- Maintenance with oxygen alone, if anesthetic plane is deep enough
- Allow the patient to breathe room air only if the patient is well oxygenated (can monitor with pulse oximeter)
- Removal of the face mask at the instant collection of radiographs
- Maintenance of supplemental isoflurane (by face mask or orotracheal intubation) during oral endoscopy
- Maintenance of supplemental isoflurane (by face mask or orotracheal intubation) during intraoral dental procedure
- Maintenance of supplemental isoflurane (by face mask or orotracheal intubation) during the extraoral procedures

RADIOLOGY

- Know your equipment, get accustomed to correct setting for rabbit skull views Mammography x-ray film provides much greater detail and high resolution

- Review different radiographic projections in class: lateral, ventrodorsal, left and right obliques, and rostrocaudal; and (optional) intraoral projection
- Normal radiographic appearance: lateral skull radiograph
- Lower incisors should occlude with peg teeth
- Lines drawn parallel to roof of maxilla and floor of mandible should converge if extended way past the film; parallel or diverging lines are abnormal
- Zig-zag line to represent occlusal surface of cheek teeth
- Strong uninterrupted line of cortical bone at base of mandible below tooth roots
- Normal radiographic appearance: VD radiograph
- First upper premolar teeth visible
- Evaluate symmetry and quality of tympanic bullae
- Evaluate zygomatic arches, orbit, etc.
- Normal radiographic appearance: oblique views
- Allows evaluation of a non-superimposed view of the cheek teeth
- Primarily evaluating the lower cheek teeth roots
- May also be able to view upper cheek teeth roots
- Normal radiographic appearance: rostrocaudal projection
- Evaluate another view of the occlusal surfaces and the resting position of the upper and lower cheek teeth
- Nasolacrimal duct can be evaluated radiographically with contrast media
- Specialty rabbit/rodent dental instrumentation
- Mouth gag cheek dilators spatula
- Dental rasp
- Table top mouth gag device (great investment!)

ORAL ENDOSCOPY

- Endoscopic equipment
- 2.7-mm rigid 30-degree bevel endoscope
- Light cable, light source

- Videocamera for endoscope recording device
- TV monitor

Why Oral Endoscopy?

Technically simple procedure

- Requires no special skill or specialization Allows a thorough inspection of the oral cavity offers a magnified perspective of dental disease Highly reduces risk of missing lesion
- Facilitates coronal reduction and other therapeutic procedures
- Can facilitate endotracheal intubation
- Allows recording of pictures for veterinarian's database and for demonstration to the owner

ADDITIONAL DIAGNOSTIC TESTING

- Culture and sensitivity
- Histopathology
- CT Scan
- Early stage of cheek teeth malocclusion
- "Wave mouth" and "step mouth"
- Spurs (lower cheek teeth usually get lingual spurs; upper cheek teeth get buccal spurs)
- Mid- to late-stage cheek teeth malocclusion
- Tongue lesions (ulcerations)
- Cheek soft tissue lesions (ulcerations)
- Root elongation
- Osteomyelitis
- Facial abscessation

4

STERILIZATION AND DISINFECTION

In 1996 in the United States, 46,500,000 surgical procedures and an even larger number of invasive medical procedures were performed . For example, ~5 million gastrointestinal endoscopies are performed per year . Each of these procedures involves contact by a medical device or surgical instrument with a patient's sterile tissue or mucous membranes. A major risk of all such procedures is the introduction of pathogenic microbes, which can lead to infection. For example, failure to properly disinfect or sterilize equipment may lead to person-to-person transmission via contaminated devices (e.g., Mycobacterium tuberculosis—contaminated bronchoscopes).

Achieving disinfection and sterilization through the use of disinfectants and sterilization practices is essential for ensuring that medical and surgical instruments do not transmit infectious pathogens to patients. Because it is not necessary to sterilize all patient-care items, health care policies must identify whether cleaning, disinfection, or sterilization is indicated, primarily on the basis of each item's intended use.

Multiple studies in many countries have documented lack of compliance with established guidelines for disinfection and sterilization. Failure to comply with scientifically based guidelines has led to numerous outbreaks of infection. In this article, a pragmatic approach to the judicious selection and proper use of disinfection and sterilization processes is presented that is based on the results of well-designed studies assessing the efficacy (via laboratory investigations) and effectiveness (via clinical studies) of disinfection and sterilization procedures.

Sterilization and disinfection are the basic components of hospital infection control activities. Every day, a number of hospitals are performing various surgical procedures. Even more number of invasive procedures are being performed in different health care facilities. The medical device or the surgical instrument that comes in contact with the sterile tissue or the mucus membrane of the patient during the various processes is associated with increased risk of introduction of pathogens into the patient's body. Moreover, there is chance of transmission of infection from patient to patient; from patient or to health care personnel, and vice versa e.g., hepatitis B virus (HBV); or from the environment to the patient (e.g., Pseudomonas aeruginosa, Acinetobacter spp.) through the improper sterilized or disinfected devices. A number of outbreaks and infections were reported in the hospital setup because of improperly sterilized devices. Many infections were reported throughout the world due to use of contaminated endoscopes. Hence, adequate decontamination techniques for medical and surgical devices are needed in all the health care facilities. The health care providers are equally responsible for the reduction and elimination of such infections. Every hospital should have its own guidelines of sterilizing and disinfecting items on the basis of their intended use of medical devices and associated infections. Currently, there is an increase in frequency of newly emerging and multidrug-resistant pathogens in all health care centers for which only few/if any treatments are available. Hence, the medical personnel, laboratory people, and health care providers should have better knowledge regarding these techniques to prevent the spread of these pathogens.

There is no uniform definition or standards for sterilization, disinfection, and cleaning. However, the Clinical and Laboratory Standards Institute has defined some minimum inhibitory concentration testing–based recommendations, which are currently standardized by the International Organization of Standardization.8 The norms were first established in 2004, and later in 2010, new guidelines were published.

Sterilization: Sterilization is defined as a process of complete elimination or destruction of all forms of microbial life (i.e., both vegetative and spore forms), which is carried out by various physical and chemical methods. Technically, there is reduction ≥106 log colony forming units (CFU) of the most resistant spores achieved at the half-time of a regular cycle.

Chemical sterilant: These are chemicals used for a longer duration (3–12 h) to destroy all forms of microbes, e.g., peracetic acid (PAA) (0.2%), glutaraldehyde (≥2.4%), ortho-phthalaldehyde (OPA) (0.55%), and hydrogen peroxide (7.5%).

Disinfection: Disinfection is defined as a process of complete elimination of vegetative forms of microorganisms except the bacterial spores from inanimate objects. Technically, there is reduction of ≥103 log CFU of microorganisms by this method without spores.

High-level disinfectant (HLD): Used for shorter duration and able to kill 106 log microorganisms except spores, e.g., glutaraldehyde (≥2.0%), OPA (0.55%), hydrogen peroxide (7.5%), hypochlorite (650–675 ppm), and hypochlorous acid (400–450 ppm).

Intermediate level disinfectant (ILD): These disinfectants act against Mycobacterium tuberculosis and are mainly used for noncritical items contaminated with blood/body fluids.

Low-level disinfectant (LLD): LLDs are used to remove the vegetative form of bacteria, few fungi, and some enveloped viruses from the noncritical items, e.g., 3% hydrogen peroxide, quaternary ammonium compound, diluted glutaraldehyde, phenolics, etc.

Decontamination and cleaning: Decontamination is the process of removal of pathogenic microorganisms from objects so that they are safe to handle. Cleaning is defined as removal of visible soil (e.g., organic and inorganic materials) from the surfaces and objects. Technically, it achieves minimum reduction of ≥1 log CFU of microorganisms.

Antisepsis: Antisepsis is a process of removal of germs from the skin. When it is related to the patient's skin, it means disinfection of living tissue or skin. When it is related to the health care worker, it means reduction or removal of transient microbe from the skin.

Germicide: It is the agent that destroys germs. It includes both antiseptics and disinfectants. The type of microorganism is identified from the prefix (e.g., virucide, fungicide, bactericide, sporicide, and tuberculocide).

A RATIONAL APPROACH TO DISINFECTION AND STERILIZATION

More than 35 years ago, Spaulding devised a rational approach to disinfection and sterilization of patient-care items or equipment. This classification scheme is so clear and logical that it has been retained, refined, and successfully used by infection-control professionals and others when planning methods for disinfection or sterilization. Spaulding believed that the nature of disinfection could be understood more readily if instruments and items for patient care were divided into 3 categories—namely, critical, semicritical, and noncritical—on the basis of the degree of risk of infection involved in the use of the items. This terminology is employed by the Centers for Disease Control and Prevention (CDC) in the documents "Guidelines for Environmental Infection Control in Health-Care Facilities" and "Guideline for Disinfection and Sterilization in Healthcare Facilities".

Critical items. Critical items are those associated with a high risk of infection if the item is contaminated with any microorganism, including bacterial spores. Thus, sterilization of objects that enter sterile tissue or the vascular system is critical, because any microbial contamination could result in disease transmission. This category includes surgical instruments, cardiac and urinary catheters, implants, and ultrasound probes used in sterile body cavities. The items in this category should be purchased as sterile or should be sterilized by steam sterilization, if possible.

If the item is heat sensitive, it may be treated with ethylene oxide (ETO) or hydrogen peroxide gas plasma or with liquid chemical sterilants if other methods are unsuitable. Tables 1 and 2 list several germicides that are categorized as chemical sterilants.

These include ≥2.4% glutaraldehyde—based formulations, 1.12% glutaraldehyde with 1.93% phenol/phenate, 7.5% stabilized hydrogen peroxide, 7.35% hydrogen peroxide with 0.23% peracetic acid, ≥0.2% peracetic acid, and 1.0% hydrogen peroxide with 0.08% peracetic acid. The indicated exposure times are within the range 3–12 h, with the exception of ≥0.2% peracetic acid (sporicidal time of 12 min at 50°C–56°C). Use of liquid chemical sterilants is a reliable method of sterilization only if cleaning precedes treatment, which eliminates organic and inorganic material, and if the proper guidelines for concentration, contact time, temperature, and pH are followed.

Another limitation to sterilization of devices with liquid chemical sterilants is that the devices cannot be wrapped during processing in the liquid chemical sterilant; thus, maintaining sterility after processing and during storage is impossible. Furthermore, after exposure to the liquid chemical sterilant, devices may require rinsing with water that, in general, is not sterile. Therefore, because of the inherent limitations of the use of liquid chemical sterilants in a nonautomated reprocessor, their use should be restricted to reprocessing critical devices that are heat sensitive and incompatible with other sterilization methods.

Process, method	Level of microbial inactivation	Example(s) (processing time)	Health care application (examples)
Sterilization			
High temperature	Destroys all microorganisms, including bacterial spores	Steam (~40 min) and dry heat (1–6 h, depending on temperature)	Heat-tolerant critical (surgical instruments) and semicritical patient-care items
Low temperature	Destroys all microorganisms, including bacterial spores	ETO gas (~15 h) and hydrogen peroxide gas plasma (~50 min)	Heat-sensitive critical and semicritical patient-care items
Liquid immersion	Destroys all microorganisms, including bacterial spores	Chemical sterilants [a] ≥2.4% glut (~10 h), 1.12% glut and 1.93% phenol (12 h), 7.35% HP and 0.23% PA (3 h), 7.5% HP (6 h), 1.0% HP and 0.08% PA (8 h), and ≥0.2% PA (~50 min at 50°C–56°C)	Heat-sensitive critical and semicritical patient-care items that can be immersed
High-level disinfection			
Heat automated	Destroys all microorganisms except high numbers of bacterial spores	Pasteurization (~50 min)	Heat-sensitive semicritical patient-care items (respiratory-therapy equipment)
Liquid immersion	Destroys all microorganisms except high numbers of bacterial spores	Chemical sterilants or high-level disinfectants [a] >2% glut (20–45 min), 0.55% OPA (12 min), 1.12% glut and 1.93% phenol (20 min), 7.35% HP and 0.23% PA (15 min), 7.5% HP (30 min), 1.0% HP and 0.08% PA (25 min), and 650–675 ppm chlorine (10 min)	Heat-sensitive semicritical patient-care items (GI endoscopes and bronchoscopes)
Intermediate-level disinfection, liquid contact	Destroys vegetative bacteria, mycobacteria, most viruses, and most fungi but not bacterial spores	EPA-registered hospital disinfectants with label claiming tuberculocidal activity, such as chlorine-based products and phenolics (at least 60 s)	Noncritical patient-care items (blood-pressure cuff) or surfaces (bedside table), with visible blood
Low-level disinfection, liquid contact	Destroys vegetative bacteria and some fungi and viruses but not mycobacteria or spores	EPA-registered hospital disinfectants with no tuberculocidal claim, such as chlorine-based products, phenolics, and quaternary ammonium compounds (at least 60 s), or 70%–90% alcohol	Noncritical patient-care items (blood-pressure cuff) or surfaces (bedside table), with no visible blood

NOTE. Modified from [13], [14], and [17]. AER, automated endoscope reprocessing; EPA, Environmental Protection Agency; ETO, ethylene oxide; FDA, US Food and Drug Administration; GI, gastrointestinal; glut, glutaraldehyde; HP, hydrogen peroxide; PA, peracetic acid; OPA, ortho-phthalaldehyde.

[a] Consult FDA-cleared package inserts for information about FDA-cleared contact time and temperature; see text for discussion of why one product (2% glut) is used at reduced exposure (20 min at 20°C). Increasing the temperature by using AER will reduce the contact time (e.g., for OPA, 12 min at 20°C, but 5 min at 25°C in AER). Tubing must be completely filled for high-level disinfection and liquid chemical sterilization. Compatibility of material should be investigated when appropriate (e.g., HP and HP with PA will cause functional damage to endoscopes).

Table 4.1: Methods for disinfection and sterilization of patient-care items and environmental surfaces.

Sterilization method	Advantages	Disadvantages
Peracetic acid and hydrogen peroxide	No activation required Odor or irritation not significant	Concerns regarding compatibility with materials (lead, brass, copper, and zinc) and both cosmetic and functional damage Limited clinical use Potential for eye and skin damage
Glutaraldehyde	Numerous published studies of use Relatively inexpensive Excellent compatibility with materials	Respiratory irritation from glutaraldehyde vapor Pungent and irritating odor Relatively slow mycobactericidal activity Coagulates blood and fixes tissue to surfaces Allergic contact dermatitis
Hydrogen peroxide	No activation required May enhance removal of organic material and organisms No disposal issues No odor or irritation issues Does not coagulate blood or fix tissues to surfaces Inactivates *Cryptosporidium* Published studies of use	Concerns regarding compatibility with materials (brass, zinc, copper, and nickel/silver plating) and both cosmetic and functional damage Serious eye damage with contact
Ortho-phthalaldehyde	Fast-acting high-level disinfectant No activation required Odor not significant Claim of excellent compatibility with materials Claim of not coagulating blood or fixing tissues to surfaces	Stains protein gray (e.g., skin, mucous membranes, clothing, and environmental surfaces) Limited clinical use More expensive than glutaraldehyde Eye irritation with contact Slow sporicidal activity Repeated exposure may result in hypersensitivity in some patients with bladder cancer
Peracetic acid	Rapid sterilization cycle time (30–45 min) Low-temperature (50°C–55°C) liquid-immersion sterilization Environmentally friendly by-products (acetic acid, O_2, and H_2O) Fully automated Single-use system eliminates need for concentration testing Standardized cycle May enhance removal of organic material and endotoxin No adverse health effects to operators, under normal operating conditions Compatible with many materials and instruments Does not coagulate blood or fix tissues to surfaces Sterilant flows through scope, facilitating salt, protein, and microbe removal Rapidly sporicidal Provides procedure standardization (constant dilution, perfusion of channel, temperatures, and exposure)	Potential incompatibility with materials (e.g., aluminum anodized coating becomes dull) Used for immersible instruments only Biological indicator may not be suitable for routine monitoring Only one scope or a small number of instruments can be processed in a cycle More expensive (endoscope repairs, operating costs, and purchase costs) than high-level disinfection Serious eye and skin damage (concentrated solution) with contact Point-of-use system; no sterile storage

NOTE. Modified from [18]. All products are effective in the presence of organic soil, are relatively easy to use, and have a broad spectrum of antimicrobial activity (bacteria, fungi, viruses, bacterial spores, and mycobacteria). The above characteristics are documented in the literature; contact the manufacturer of the instrument and sterilant for additional information. All products listed have been cleared by the US Food and Drug Administration (FDA) as chemical sterilants except for ortho-phthalaldehyde, which is an FDA-cleared high-level disinfectant.

Table 4.2: Summary of advantages and disadvantages of chemical agents used as chemical sterilants or as high-level disinfectants.

Semicritical items. Semicritical items are those that come in contact with mucous membranes or nonintact skin. Respiratory-therapy and anesthesia equipment, some endoscopes, laryngoscope blades, esophageal manometry probes, anorectal manometry catheters, and diaphragm-fitting rings are included in this category. These medical devices should be free of all microorganisms (i.e., mycobacteria, fungi, viruses, and bacteria), although small numbers of bacterial spores may be present. In general, intact mucous membranes, such as those of the lungs or the gastrointestinal tract, are resistant to infection by common bacterial spores but are susceptible to other organisms, such as bacteria, mycobacteria, and viruses. The minimum requirement for semicritical items is high-level disinfection using chemical disinfectants. Glutaraldehyde, hydrogen peroxide, ortho-phthalaldehyde (OPA), peracetic acid with hydrogen peroxide, and chlorine have been cleared by the US Food and Drug Administration (FDA) and are dependable high-level disinfectants when guidelines for effective germicidal procedures are followed (tables 1 and 2). The exposure time for most high-level disinfectants varies from 10 to 45 min, at 20°C–25°C. Outbreaks of infection continue to occur when ineffective disinfectants, including iodophor, alcohol, and over-diluted glutaraldehyde , are used for so-called high-level disinfection. When a disinfectant is selected for use with certain patient-care items, the chemical compatibility after extended use with the items to be disinfected must also be considered. For example, compatibility testing by Olympus America of 7.5% hydrogen peroxide showed cosmetic and functional changes in the tested endoscopes (Olympus America, personal communication). Similarly, Olympus America does not endorse the use of products containing hydrogen peroxide with peracetic acid, because of cosmetic and functional damage (Olympus America, personal communication).

Semicritical items that will have contact with the mucous membranes of the respiratory or gastrointestinal tract should be rinsed with sterile water, filtered water, or tap water, followed by an alcohol rinse. An alcohol rinse and forced-air drying markedly reduces the likelihood of contamination of the instrument (e.g., endoscopes), most likely by eliminating the wet environment favorable to bacterial growth . After rinsing, items should be dried and then stored in a manner that protects them from damage or contamination. There is no recommendation to use sterile or filtered water, rather than tap water, for rinsing semicritical equipment that will have contact with the mucous membranes of the rectum (e.g., rectal probes or anoscopes) or vagina (e.g., vaginal probes) .

Noncritical items. Noncritical items are those that come in contact with intact skin but not mucous membranes. Intact skin acts as an effective barrier to most microorganisms; therefore, the sterility of items coming in contact with intact skin is "not critical." Examples of noncritical items are bedpans, blood-pressure cuffs, crutches, bed rails, linens, bedside tables, patient furniture, and floors. In contrast to critical and some semicritical items, most noncritical reusable items may be decontaminated where they are used and do not need to be transported to a central processing area. There is virtually no documented risk of transmitting infectious agents to patients via noncritical items when they are used as noncritical items and do not contact nonintact skin and/or mucous membranes. However, these items (e.g., bedside tables or bed rails) could potentially contribute to secondary transmission, by contaminating the hands of health care workers or by contact with medical equipment that will subsequently come in contact with patients . Table 1 lists several low-level disinfectants that may be used for noncritical items. The exposure times for these disinfectants are 60 s or longer.

CURRENT ISSUES IN DISINFECTION AND STERILIZATION

Reprocessing of endoscopes. Physicians use endoscopes to diagnose and treat numerous medical disorders. Although endoscopes are a valuable diagnostic and therapeutic tool in modern medicine and although the incidence of infection associated with their use has been reported to be very low , more health care—associated outbreaks of infection have been linked to contaminated endoscopes than to any other medical device. To prevent the spread of health care—associated infection, all heat-sensitive endoscopes (e.g., gastrointestinal endoscopes, bronchoscopes, and nasopharyngoscopes) must be properly cleaned and, at a minimum, subjected to high-level disinfection after each use. High-level disinfection can be expected to destroy all microorganisms, although a few bacterial spores may survive when high numbers of spores are present.

Recommendations for the cleaning and disinfection of endoscopic equipment have been published and should be strictly followed. Unfortunately, audits have shown that personnel do not adhere to guidelines on reprocessing and that outbreaks of infection continue to occur. To ensure that the personnel responsible for reprocessing are properly trained, initial and annual competency testing should be required for each individual who is involved in reprocessing endoscopic instruments.

In general, endoscope disinfection or sterilization with a liquid chemical sterilant or high-level disinfectant involves the following 5 steps, which should be performed after leak testing: (1) clean: mechanically clean internal and external surfaces, including brushing internal channels and flushing each internal channel with water and an enzymatic cleaner; (2) disinfect: immerse endoscope in high-level disinfectant (or chemical sterilant), perfuse disinfectant (which eliminates air pockets and ensures contact of the germicide with the internal channels) into all accessible channels, such as the suction/biopsy channel and the air/water channel, and expose endoscope for the time recommended for specific products; (3) rinse: rinse the endoscope and all channels with sterile water, filtered water (commonly used with automated endoscope reprocessors), or tap water; (4) dry: rinse the insertion tube and inner channels with alcohol and dry with forced air, after disinfection and before storage; and (5) store: store the endoscope in a way that prevents recontamination and promotes drying (e.g., hung vertically).

Unfortunately, there is poor compliance with the recommendations for reprocessing endoscopes. In addition, in rare instances, the scientific literature and recommendations from professional organizations regarding the use of disinfectants and sterilants may differ from claims on the manufacturer's label. One example is the contact time used to achieve high-level disinfection with 2% glutaraldehyde. On the basis of FDA requirements (the FDA regulates liquid sterilants and high-level disinfectants used on critical and semicritical medical devices), manufacturers test the efficacy of their germicide formulations under worst-case conditions (i.e., minimum recommended concentration of the active ingredient) and in the presence of organic soil (typically, 5% serum). The soil represents the organic loading to which the device is exposed during actual use and that would remain on the device in the absence of cleaning. These stringent test conditions are designed to provide a margin of safety, by assuring that the contact conditions for the germicide provide complete elimination of the test bacteria (e.g., 105–106 cfu M. tuberculosis in organic soil and dried on a scope) if inoculated into the most difficult areas for the disinfectant to penetrate and in the absence of cleaning. However, the scientific data demonstrate that M. tuberculosis levels can be reduced by at least 8 log10 cfu with cleaning (reduction of 4 log10 cfu) followed by chemical disinfection for 20 min at 20°C (reduction of 4–6 log10 cfu). Because of these data, professional organizations (at least 14 worldwide) that have endorsed an endoscope-reprocessing guideline recommend contact with 2% glutaraldehyde for 20 min (or <20 min outside the United States) at 20°C to achieve high-level disinfection, which differs from the recommendation given on the manufacturer's label.

It is important to emphasize that the FDA tests do not include cleaning, a critical component of the disinfection process. When cleaning has been included in the test methodology, contact with 2% glutaraldehyde for 20 min has been demonstrated to be effective in eliminating all vegetative bacteria.

Inactivation of Creutzfeldt-Jakob disease (CJD) agent. CJD is a degenerative neurologic disorder in humans, with an incidence in the United States of ~1 case/million population/year [35]. CJD is thought to be caused by a proteinaceous infectious agent, or prion. CJD is related to other human transmissible spongiform encephalopathies (TSEs), such as kuru (now eradicated), Gertsmann-Straussler-Sheinker syndrome (1 case/40 million population/year), and fatal insomnia syndrome (<1 case/40 million population/year). The agents of CJD and other TSEs exhibit an unusual resistance to conventional chemical and physical decontamination methods. Because the CJD agent is not readily inactivated by conventional disinfection and sterilization procedures and because of the invariably fatal outcome of CJD, the procedures for disinfection and sterilization of the CJD prion have been both conservative and controversial for many years.

The current recommendations consider inactivation data but also use epidemiological studies of prion transmission, infectivity of human tissues, and efficacy of removing proteins by cleaning. On the basis of scientific data, only critical devices (e.g., surgical instruments) and semicritical devices contaminated with high-risk tissue (i.e., brain, spinal cord, or eye tissue) from high-risk patients (e.g., known or suspected infection with CJD or other prion disease) require special prion reprocessing. When high-risk tissues, high-risk patients, and critical or semicritical medical devices are involved, one of the following methods should be used: cleaning of the device and sterilization using a combination of sodium hydroxide and autoclaving (e.g., immerse in 1N NaOH for 1 h, remove and rinse in water, and then transfer to an open pan for autoclaving for 1 h at 121°C in a gravity displacement sterilizer or at 134°C in a porous or prevacuum sterilizer); autoclaving for 18 min at 134°C in a prevacuum sterilizer; or autoclaving for 1 h at 132°C in a gravity displacement sterilizer). The temperature should not exceed 134°C, because the effectiveness of autoclaving may decline as the temperature is increased (e.g., to 136°C or 138°C). Prion-contaminated medical devices that are impossible or difficult to clean should be discarded. Flash sterilization (i.e., steam sterilization of an unwrapped item for 3 min at 132°C) should not be used for reprocessing. To minimize environmental contamination, noncritical environmental surfaces should be covered with plastic-backed paper; when contaminated with high-risk tissues, the paper should be properly discarded. Noncritical environmental surfaces (e.g., laboratory surfaces) contaminated with high-risk tissues should be cleaned and then spot decontaminated with a 1:10 dilution of hypochlorite solution.

Emerging pathogens, antibiotic-resistant bacteria, and bioterrorism agents. Emerging pathogens are of growing concern to the general public and infection-control professionals. Relevant pathogens include Cryptosporidium parvum, Helicobacter pylori, Escherichia coli O157:H7, HIV, hepatitis C virus, rotavirus, multidrug-resistant M. tuberculosis, human papillomavirus, and nontuberculosis mycobacteria (e.g., Mycobacterium chelonae). Similarly, recent publications have highlighted concern about the potential for biological terrorism. The CDC has categorized several agents as "high priority" because they can be easily disseminated or transmitted by person-to-person contact, can cause high mortality, and are likely to cause public panic and social disruption. These agents include Bacillus anthracis (anthrax), Yersinia pestis (plague), variola major (smallpox), Francisella tularensis (tularemia), filoviruses (Ebola and Marburg [hemorrhagic fever]), and arenaviruses (Lassa [Lassa fever] and Junin [Argentine hemorrhagic fever]) and related viruses.

With rare exceptions (e.g., human papillomavirus), the susceptibility of each of these pathogens to chemical disinfectants or sterilants has been studied, and all of these pathogens (or surrogate microbes, such as feline calicivirus for Norwalk virus, vaccinia for variola, and Bacillus atrophaeus [formerly Bacillus subtilis] for B. anthracis) have been found to be susceptible to currently available chemical disinfectants or sterilants. Standard sterilization and disinfection procedures for patient-care equipment (as recommended in this article) are adequate for sterilization or disinfection of instruments or devices contaminated with blood or other body fluids from persons infected with bloodborne pathogens, emerging pathogens, or bioterrorism agents, with the exception of prions (see previous section). No changes in procedures for cleaning, disinfecting, or sterilizing need to be made. In addition, there are no data to show that antibiotic-resistant bacteria (e.g., methicillin-resistant Staphylococcus aureus, vancomycin-resistant Enterococcus faecium, and multidrug-resistant M. tuberculosis) are less sensitive to liquid chemical germicides than are antibiotic-sensitive bacteria at currently used germicide contact conditions and concentrations.

Advances in disinfection and sterilization methods. In the past several years, new methods of disinfection and sterilization have been introduced in health care settings. OPA is a chemical sterilant that received FDA clearance in October 1999. It contains 0.55% 1,2-benzenedicarboxaldehyde. In vitro studies have demonstrated excellent microbicidal activity. For example, Gregory et al. demonstrated that OPA has shown superior mycobactericidal activity (reduction of 5 log10 in 5 min), when compared with glutaraldehyde. The advantages, disadvantages, and characteristics of OPA are listed in table 2.

The FDA recently cleared a liquid high-level disinfectant (superoxidized water) that contains 650–675 ppm free chlorine and a new sterilization system using ozone. Because there are limited data in the scientific literature for assessing the antimicrobial activity or material compatibility of these processes, they have not yet been integrated into clinical practice in the United States .

Several methods are used to sterilize patient-care items in health care, including steam sterilization, ETO, hydrogen peroxide gas plasma, and a peracetic acid—immersion system. The advantages and disadvantages of these systems are listed in table 3.

Sterilization method	Advantages	Disadvantages
Steam	Nontoxic to patient, staff, and environment Cycle is easy to control and monitor Rapidly microbicidal Least affected by organic/inorganic soils among sterilization processes listed Rapid cycle time Penetrates medical packing and device lumens	Deleterious for heat-sensitive instruments Microsurgical instruments damaged by repeated exposure May leave instruments wet, causing them to rust Potential for burns
Hydrogen peroxide gas plasma	Safe for the environment Leaves no toxic residuals Cycle time is 45–73 min, and no aeration is necessary Used for heat- and moisture-sensitive items, because process temperature is <50°C Equipment is simple to operate, install (208 V outlet), and monitor Compatible with most medical devices Equipment requires electrical outlet only	Cellulose (paper), linens, and liquids cannot be processed Sterilization chamber is small (~3.5–7.3 ft^3) Endoscope or medical-device restrictions based on lumen internal diameter and length (see manufacturer's recommendations) Requires synthetic packaging (polypropylene wraps or polyolefin pouches) or special container tray Hydrogen peroxide may be toxic at levels >1 ppm TWA
100% ETO	Penetrates packaging materials and device lumens Single-dose cartridge and negative-pressure chamber minimizes the potential for gas leak and ETO exposure Equipment is simple to operate and monitor Compatible with most medical materials	Requires aeration time to remove ETO residue Sterilization chamber is small (~4–8.8 ft^3) ETO is toxic, a carcinogen, and flammable ETO emission regulated by states, but catalytic cell removes 99.9% of ETO and converts it to CO_2 and H_2O ETO cartridges should be stored in flammable liquid-storage cabinet Lengthy cycle and aeration time
ETO mixture[a]	Penetrates medical packaging and many plastics Compatible with most medical materials Cycle is easy to control and monitor	Some states (e.g., CA, NY, and MI) require ETO-emission reduction of 90%–99.9% CFC (inert gas that eliminates explosion hazard) banned in 1995 Potential hazards to staff and patients Lengthy cycle and aeration time ETO is toxic, a carcinogen, and flammable
Peracetic acid	Rapid cycle time (30–45 min) Low-temperature (50°C–55°C) liquid immersion sterilization Environmentally friendly by-products Sterilant flows through endoscope, which facilitates salt, protein, and microbe removal	Point-of-use system, no sterile storage Biological indicator may not be suitable for routine monitoring Used for immersible instruments only Some incompatibility with materials (e.g., aluminum anodized coating becomes dull) Only 1 scope or a small number of instruments processed in a cycle Potential for serious eye and skin damage (concentrated solution) with contact Must use connector between system and scope to ensure infusion of sterilant to all channels

NOTE. Modified from [46]. CFC, chlorofluorocarbon; ETO, ethylene oxide; HCFC, hydrochlorofluorocarbon; TWA, time-weighted average.
[a] 8.6% ETO and 91.4% HCFC; 10% ETO and 90% HCFC; or 8.5% ETO and 91.5% CO_2.

Table 3.3: Summary of advantages and disadvantages of commonly used sterilization technologies.

New sterilization technology based on plasma was patented in 1987 and has been marketed in the United States since 1993. Gas plasmas have been referred to as the fourth state of matter (i.e., liquid, solid, gas, and gas plasma). Gas plasmas are generated in an enclosed chamber in a deep vacuum, by using radio frequency or microwave energy to excite the gas molecules and produce charged particles, many of which are in the form of free radicals. This process has the ability to inactivate a broad spectrum of microorganisms, including resistant bacterial spores. Studies have been conducted against vegetative bacteria (including mycobacteria), yeasts, fungi, viruses, and bacterial spores . The effectiveness of all sterilization processes can be altered by lumen length, lumen diameter, inorganic salts, and organic materials.

CLASSIFICATION OF MEDICAL DEVICES AND/OR EQUIPMENT

The principal goal of cleaning, disinfection, and sterilization is to reduce the number of microorganisms on the device to such a level that the probability of transmission of infection will be nil. The risk of transmission of infection depends on the type of device. For example, in case of blood pressure–measuring cuff, the device that comes in contact with the skin carries least risk of transmission of disease. On the other hand, devices that come in contact with the neural tissue of a patient suffering from Creutzfeldt–Jakob disease (CJD) have high risk of transmission of infection. To reduce the potential risks of transmission of infection through different devices, Earle H. Spaulding in 1968 proposed a classification to define the desired level of antimicrobial killing for different devices. He categorized the devices into three categories depending on the potential risk of transmission of infectious agents: critical, semicritical, and noncritical.

This simple classification of Spaulding's needs to be revised, as it did not address the devices that come in contact with the mucous membrane (e.g., endoscope), biopsy forceps touching the breach sterile tissue, heat-sensitive items, and prions. The Centers for Disease Control and Prevention (CDC) in 1991 proposed an additional category to Spaulding's classification as "environmental surfaces" to represent the surfaces that usually do not come in contact with patient.13 Environmental surfaces can be further subgrouped as clinical contact surfaces (medical equipment or high-touch surfaces) and housekeeping surfaces. CDC defines clinical contact surfaces as the areas that act like reservoirs of microorganisms, e.g., hands of health care workers. High-touch surfaces such as telephone, light switch board, bedrails, computer, door handle, and medical equipment like ventilator, X-ray machines, and hemodialysis machines are the contacting equipment that subsequently contact the patient. CDC had issued guidelines for hand washing and hospital environmental control. Different LLDs and ILDs that can be used to disinfect the clinical contact surfaces were approved by the Environmental Protection Agency (EPA). The housekeeping surfaces such as walls, floor, and sinks carry very low risk of transmission of infection. So, disinfection of such surfaces is less frequent in comparison to the previous one.

Sterilization, Disinfection, and Cleaning in the Health Care Facilities

Sterilization, disinfection, and cleaning in health care facilities include disinfection and cleaning of environmental surfaces with/without cleaning and reprocessing the medical equipment. The former includes mainly the noncritical items such as surfaces, floors, and high-contact surfaces (sinks, telephones, switches board, bed railings, trolleys etc.). It is observed that regular cleaning of all these housekeeping surfaces dramatically reduces the transmission of the infection.

- Recommendation for the Cleaning and Decontamination of Environmental Surfaces
- The space for cleaning and other work should be clearly demarcated and separated by walls.
- The hospital staffs should be properly trained regarding the cleaning and decontamination practices of hospital surfaces.
- The staffs should wear personal prophylactic equipment (PPE), i.e., gowns, gloves, masks, and boots. There must be separate area for removing PPE.
- Fresh preparation of detergents or disinfectants should be made every day and used with the appropriate dilution as per the manufacturer instruction.
- Wet mopping of floors should be encouraged as dry mopping generates dust aerosols.

Avoid using HLD for environmental surfaces of the hospital and offices. Mopping of the hospital surfaces should be done using detergent. Table tops and counters should also be cleaned regularly by detergent only.

Mopping of high-risk areas such as the intensive care units (ICUs), burn wards, transplant units, isolation wards, operation theaters (OTs), and dialysis machines should be done using HLD instead of detergent. Cleaning by vacuum pump and use of high-efficiency particulate air (HEPA) filters for the exhaust are preferred in these places.

Recommendation for Blood Spill on the Surface

Cleaning and decontamination of the hospital surfaces having spilled blood is done as per the recommendation of Occupational Safety and Health Administration/ World Health Organization/CDC.

ILD or disinfectants with tuberculocidal activity should be used for blood spill in the hospital surfaces.

For decontamination of small amount of blood spills (<10 mL), sodium hypochlorite solution is used with a dilution of 1:100. For spill >10 mL, sodium hypochlorite with 1:10 dilution is used for the first application. The organic matter should be cleaned with absorbent material, and final disinfection may be done using sodium hypochlorite solution with 1:100 dilution.

Cleaning and Disinfection of Medical Instruments

The cleaning and disinfection of medical equipment depends on their physical nature, character of the material it is made up of, lumen size, etc. Thorough cleaning is preferred before the use of the disinfectants as cleaning effectively removes majority of the microbes from the equipment.

Staffs should be properly educated and trained regarding the cleaning procedure, physical and chemical nature of the instruments, nature of disinfectants, etc. All the staffs during the process should use PPE.

Dry organic materials are difficult to remove from the instrument. Hence, drying should be avoided by immersing the equipment in the detergent or disinfectant solution prior to cleaning. The soaked matter can be cleaned by manual scrubbing and rubbing with brush or automated scrubber and thoroughly washed with water under pressure. Avoid prolonged or overnight soaking of the devices.

The time of exposure, and concentration of the detergent or disinfectant, should be properly maintained as mentioned in the literature. Too low concentration may not work effectively to remove the organic materials or microorganisms.

The pH of the disinfectant should be properly obtained as per the manufacturer's instruction. Delicate articles should be processed in neutral pH.

Enzymes like proteases may be added to the solution to fasten the cleaning action. Enzymatic cleaners with neutral pH are preferred to avoid the damage of the articles. For example, in case of flexible endoscope, neutral pH detergent with enzymatic action is preferred. A new nonenzyme product [hydrogen peroxide based, US Food and Drug Administration (FDA) cleared] has been found to be very effective as cleaning agent.

Reprocessing of the Critical Items

Devices that come in contact with the sterile parts of the body are included in critical items category. They carry the highest risk of transmission of infection. Hence, sterilization is the method of choice for the reprocessing of these items (heat stable). The FDA has approved ethylene oxide (EtO), plasma sterilization, and liquid sterilization with glutaraldehyde or PAA in heat-sensitive items.22, 23 All packed sterile items should be kept with proper precaution to avoid environmental contamination.

Reprocessing of Semicritical Items

Items that come in contact with the mucous membrane of the skin are included in this category. These items should be processed by either heat sterilization or HLD after cleaning.

All the semicritical items should be rinsed with sterile water or alcohol. Forced air drying after the rinsing process drastically reduces the rate of contamination. It is found that cleaning also reduces the transmission of infection in human immunodeficiency virus (HIV)–contaminated instruments. Items are found to be germ free when soaked in 2% glutaraldehyde for 20 min after the cleaning process. OPA, glutaraldehyde, and automated process using PAA are the three disinfectants commonly used for the reprocessing of endoscopes.

Reprocessing of Noncritical Items

Items that come in contact with the intact skin are included within noncritical items. These include clothing, floors, high-touch surfaces, furniture, baths, bed pans, weighing scale, brushes, beddings, crockery, earphones, mobiles, and trolleys. The risk of transmission of infection with these items is observed to be the lowest. However, they contribute to the transmission of infection in indirect way. For example, methicillin-resistant Staphylococcus aureus (MRSA), and vancomycin-resistant enterococci (VRE) are commonly isolated with the patient's belonging and can be easily transmitted to other patient by health care worker's hand causing infection. These items do not need sterilization; however, they should be regularly cleaned and disinfected with LLD to decrease the transmission of infective organisms.

Reprocessing of Respiratory Apparatus and Endoscopes

Respiratory apparatus such as ventilators, humidifiers, nebulizers, pulmonary screening devices, anesthetic equipment, laryngoscope and its blade, and suction equipment are most important in the ICU setup because of its association with the risk of transmission of infection. Proper cleaning and infection preventive measures should be followed while handling these instruments as they are highly associated with the transmission of infection from one patient to other. They come in contact with the mucous membrane of the body, are included in the semicritical item category, and are sterilized or disinfected with the HLD.

Ventilators

Ventilators are important sources of hospital-acquired infection. This artificial airway is associated with increased chance of aspiration of the bacteria causing infection. Mechanical ventilators are directly not associated with the infection, but their internal circuits (includes the filter, tubing, humidifier, etc.) and the fluids are the potential source of infection. As per the CDC guidelines, the permanent circuits should be replaced with sterile ones, when there is visible soiling or mechanical obstruction. It has been also seen that changing interval of tubing at 7, 14, and 30 days drastically reduces the transmission of infection. In case of detachable circuits, it should be dismantled, cleaned, and disinfected.

Routine Ventilator Care

- Clean the equipment regularly and cover the ventilator, when it is not in use.
- Use sterile water to fill the humidifier as tap water causes introduction of microorganisms like Burkholderia cepacia and Legionella spp.
- Use PPE, and mask during handling these equipment. Discard all disposals and perform hand hygiene after each handling.
- Tubings of the ventilator are infected with the secretion of the patient. The condensate from the inspiratory lines may spill to the tracheobronchial tree of the patient or into the nebulizer while handling, changing, or manipulating the ventilator circuit.
- The effluent from the ventilator may contaminate the environment and can reenter through ventilator to the patient's airway increasing the chance of infection.
- Select HEPA filters for both the inspiratory and expiratory limbs of ventilator circuits.
- Do not allow the condensate to drain back into the patient's airway or back to into humidifiers.
- Change the disposable parts of the ventilator after each use, and decontamination should be done after 48 h for the reusable items.
- Clean the visible soiling; sterilize the parts with autoclaving/low-temperature sterilization.
- Infant ventilators should be sterilized with EtO.
- Toxic residues should be removed after each cycle of sterilization by flushing with air and oxygen.

Humidifiers

For proper humidification, use sterile water in place of tap water to fill the humidifiers. The fluid should be dispensed aseptically without entering or touching by hand. On the other hand, heat-moisture exchanger (HME) can be used. It absorbs the heat and humidity from the expired air of the patient and stores it. During the inhalation process, the cold dry gas entering to the ventilator absorbs this heat and moisture, thereby reducing the formation of condensate. HME should be exchanged in case of gross contamination, mechanical dysfunction, or in between patients. Humidifiers can be cleaned using 70–90% alcohol. No antiseptic should be added to the water used for the humidifier.

Nebulizer

Use sterile water for nebulization. The remaining fluid or medication should be handled aseptically. After every use, the nebulizer cap should be made dry or flushed with 70–90% alcohol before filling with water. The mouthpiece and mask should also be cleaned with warm water and dried before every use.

ANESTHETIC EQUIPMENT

Anesthetic equipment such as face mask, ambu bag, tubings, and endotracheal tubes should be regularly cleaned. In patients suspected of tuberculosis, disposable face mask and tubings should be used. Ambu bag should be kept covered to avoid exposure to dust particles. The bags should be changed in case of visible soiling or secretion.

Pulmonary Screening Device

Do not routinely sterilize or disinfect the internal machinery of pulmonary function test machines. They should be wiped and disinfected with HLD in between patients. All the screening devices (inspiratory force manometer, tidal volume/vital capacity devices, and peak flow meters) should be discarded after single use.

Reprocessing of Endoscopes

Endoscopes are very useful tools for diagnostic as well as therapeutic processes. Many outbreaks are reported with the contaminated endoscopes due to faulty reprocessing processes. Reprocessing of endoscopes remains the most challenging task in the health care facilities. Although the incidence of infection associated with the contaminated endoscope is quite low (~1 in 1.8 million procedures), its frequency is found higher in comparison to other devices. The bioburden depends on the body cavity it is intended to visualize. For example, endoscopes used for gastrointestinal tract harbor 105 CFU–1010 CFU/mL compared with 6.4 × 104 CFU/mL in bronchoscopes. It also depends on the material it is made up of. Most of the flexible endoscopes (e.g., flexible bronchoscope, gastroscope, duodenoscope, sigmoidoscopes) are found to be heat sensitive and unable to withstand many chemicals. Flexible endoscopes have multiple channels, small lumen, as well as honeycombed and blind ends, which are very difficult to clean. The endoscopes can acquire contamination from the patients, hospital environment, or water supply. It has been noticed that procedures such as endoscopic retrograde cholangiopancreatography (ERCP) are associated with many iatrogenic infection.

When the endoscopes touch the sterile tissue, they are classified as critical items and sterilization or HLD is the ideal procedure for reprocessing. Endoscopes coming in contact with the mucus membrane are classified as semicritical items, and HLD should be used for the reprocessing. With proper cleaning, the level of bioburden seems to be decreased by 4–6 log10. Many reports showed that with proper cleaning decontamination could be achieved from HIV also. Few advice the application of 20% glutaraldehyde for 20 min after the cleaning step. The FDA has approved a list of HLDs/sterilants for reprocessing endoscopes. The following norms should be followed by the staffs to reduce the faults during reprocessing.

Each person who reprocesses should be properly trained, use PPE (gowns, gloves, goggles, face mask, etc.) during the process

They should be properly educated about the chemical and biological hazards and comply with the manufacturer's instruction.

Determine which process of sterilization is suitable for your endoscope. If it is heat stable, use autoclave/low-temperature sterilization methods such as hydrogen peroxide gas plasma or PAA, whereas liquid sterilants/HLDs can be used for heat-sensitive scopes.

In general, reprocessing of the scopes includes five steps after a leak testing: cleaning, disinfection, rinsing, drying, and storing.

Cleaning

Cleaning can be done by completely immersing the scopes in the detergent/disinfectant solution. All the components should be dismantled before immersion. The internal and external surface of the scopes should be gently cleaned with the help of brush or soft cloth. Brushes should be applied to the orifices and internal surfaces to remove the organic residues. All the items should be cleaned thoroughly, and the disinfectant/detergent should be discarded after cleaning. Determine whether the scope can be suitable for automatic washer with sterilizers. These systems contain PAA/hydrogen peroxide plasma-based (HPP), which are highly effective for killing vegetative and spore forms. Hence, precleaning step is not required with these systems.

Disinfection

First, the endoscope and its accessories should be completely immersed into the disinfectant/sterilant. All the channels should be profused with the disinfectant. The air pockets should be removed for adequate contact of the sterilant with surfaces. The time of exposure, concentration, and temperature should be maintained as per the instruction. The disinfectants that should not be used for endoscopes are chlorine compounds, iodoforms, quaternary ammonium compounds, phenols, and alcohols. OPA for 12 min was found to be more advantageous than glutaraldehyde. Low-temperature sterilization can be achieved with EtO, but it is very lengthy, toxic, and expensive.

Rinse and Dry

After the treatment with the sterilant, the endoscopes should be thoroughly washed with sterile water/filtered water to remove all the residual chemicals. The scopes should be dried flushing 70–90% alcohol and forced air. Drying should be done after disinfection and before storage to reduce the contamination.

Store

The disinfected endoscopes should be dried, capped, and kept vertically for the prevention of contamination. If automated washer is used, regular maintenance and disinfection of automated washer should be done. Protocol should be developed to know whether the endoscopes are properly cleaned and disinfected or not. After the reprocessing, it can be used up to for 1/2 weeks with the maintenance of proper storage condition.

RISK OF NONENDOSCOPIC TRANSMISSION OF INFECTION

There are many issues regarding nonendoscopic transmission of various infections. There were many outbreaks due to the mishandling of various parts of endoscopes [e.g., intravenous (IV) tubings, needles, or syringes]. There is no consensus on the microbiological finding especially when the numbers of isolated bacteria are very small/isolation of environmental contaminants like coagulase-negative Staphylococcus aureus or Bacillus sp. Alfa et al. have given <100 CFU/mL as the benchmark of a reprocessed endoscope channel. It has been also observed that pathogens like M. tuberculosis and P. aeruginosa are increasingly resistant to the commonly used HLD, i.e., glutaraldehyde. Now many rapid test methods such a strip detecting the residual protein, hemoglobin carbohydrate within the channels, and bioluminescence-based tests are available. Along with the new technology of HLDs such as improved hydrogen peroxide, improved automated endoscopic reprocessor are also introduced, e.g., EVOTECH (FDA cleared). The EVOTECH machine reduces the manual cleaning process. All the scopes are put inside it after cleaning, and it checks all the leaks, disinfects, and promotes drying by alcohol flush.

Processing of Other Scopes

Laparoscope is one of the popular equipment that enters into the sterile space. It is rigid, easy to clean, and disinfect. Hence the risk of transmission of infection is also found less. It is disinfected with the HLDs. The gastrointestinal endoscopes such as duodenoscopes are used in the ERCP process. They are commonly found to be associated with introduction of iatrogenic infection. All the gastrointestinal endoscopes must be reprocessed after every patient because of high bioburden. Since majority of them are heat sensitive, they should be sterilized with HPP or PAA. Similarly, bronchoscope should be reprocessed depending upon the material used. In general, the flexible bronchoscopes are sterilized with HPP or PAA, and rigid bronchoscopes are sterilized with steam sterilization. All other flexible scopes such as the thoracoscope, laryngoscope, sinuscope, and esophagoscope are sterilized with HLDs.

Specific Issues

Cleaning and Disinfecting Surface and Floors

In general, routine cleaning and disinfection of the hospital surface and floors should be done using EPA-registered hospital detergent/disinfectant. These procedures are very important in places like the ICU, OTs, and emergency rooms, where there is likely chance of getting contamination with the blood/other body fluids. Routine environmental disinfection remains crucial particularly in case of respiratory syncytial virus infection, severe acute respiratory syndrome complex, norovirus infection, clusters of Clostridium difficile or infection with MRSA. Disinfection should be done daily for all areas with patients in contact isolation (patients infected with MRSA). Norovirus infection is highly contagious, hence items contaminated with norovirus infection vomit should be immediately disinfected with concentrated bleach or oxygen-related compound is needed.

Inactivation of Emerging/Resistant Bacteria

Emerging pathogens and multidrug-resistant bacteria are one of the major concerns in medical facilities. Emergence of antibiotic resistance in certain bacteria, e.g., multidrug-resistant M. tuberculosis, MRSA, VRE, extended spectrum of β-lactamases, and metallo-β-lactamase producers are also important causes of various infection. Except prions, most of the aforementioned pathogens are susceptible to different sterilization and disinfection procedures. Hence, standard sterilization methods and EPA-registered disinfectants should be used for all the equipment, devices, and surfaces irrespective of the patient status (known to be infected/not infected) to prevent the transmission of these infections. A sporicidal germicide such as glutaraldehyde/OPA/PAA/glutaraldehyde with phenol/H2O2 is required for the proper elimination of C. difficile spores.

Disinfection of Human Immunodeficiency Virus, Hepatitis B Virus, Hepatitis C Virus, and Tuberculosis-Contaminated Devices

All the equipment, devices, and surfaces should be treated in the same way as infected regardless of the whether the patient is known to be infected with HIV, HBV, HCV, or M. tuberculosis. These should be sterilized or disinfected with HLDs approved by the EPA or FDA.

Among the LLDs, isopropyl alcohol is one of the common disinfectants used for small surface areas at a concentration of 60–95%. It has no action on spores, but it is active against bacteria, viruses, fungi, and M. tuberculosis. Iodophore and phenolic compounds are ILDs, which are used to disinfect various surgical and medical instruments. The exposure time is about 10 min. They were found to be inactive against bacterial spores.

Special Precaution for Inactivation of Creutzfeldt–Jakob Disease

CJD has become a major crisis for the medical field because of lack of curative treatment. CJD is a highly infectious disease caused by the infection by prions (altered protein fragment). High-risk tissues include the neurological tissues such as the brain, spinal cord, and eye (retina, optic nerve). Prions are resistant to the routine sterilization methods (includes heat sterilization and chemical sterilants). There are special guidelines for the sterilization and decontamination of the instruments exposed to patients with CJD.

- Risk assessment of the patient is done with the following criteria.
- Known or suspected case of CJD or other transmissible spongiform encephalopathy
- Unusual progressive dementia with myoclonus and ataxia
- Family history of CJD/Gerstmann–Sträussler–Scheinker disease/fatal familial insomnia, dura matter transplant

- Cadaver-derived pituitary hormone injection, etc.

All the critical and semicritical devices exposed to the aforementioned surfaces are considered as infective and should be disinfected with proper care. Sodium hydroxide (1N NaOH, 121°C for 30 min) along with steam sterilization is used. Instruments that do not tolerate this above method, must be cleaned twice, treated with various chemicals such as PAA, iodophors, 3% sodium dodecyl sulfate, or 6M urea and autoclaved at 121°C for 30 min. The noncritical items are difficult to clean and should be discarded.

Health Care–Associated Infections

Definition

Hospital-acquired infection or nosocomial infection (NI) is defined as the localized or systemic infection acquired during the hospital care due to the adverse reactions to the presence of an infectious agent(s) or its toxin, which was neither present nor in incubating period during the time of admission. On the other hand, it can be defined as the infection that appeared after 48 h of admission to the hospital. Over 1.4 million people acquire infection from the hospital worldwide. In developed countries, the incidence of NI varies from 5% to 25%. The incidence of NI in pediatric ICU is estimated to be about 6.1–29.6%. Across the world, the incidence of NI (1%) was found lowest in the Netherlands.

Nowadays, health care–associated infection (HAI) has become a major arm of NI, which indirectly reflects the colonized pathogens or exposure to variety of infectious agent through contaminated devices. The infection is usually mild, but sometimes it may become severe and life threatening. Majority of infection occur in the ICUs, nurseries, and surgical units. Additionally, with the increase in the length of stay, the treatment procedures become very expensive. In Europe, HAIs contribute approximately 16 million extra stays in the hospital and 37,000 deaths per year.47 The incidence of HAI varies from 5.1% to 11.6% in developing countries in comparison to 7.1% in Europe.47 Death rate varies from infection to infection. For example, HAI due to blood stream infection and ventilator-associated pneumonia increase the death rate as well as the extra cost due to hospital stay. It was estimated that death rate due to HAI is more than death rate due to AIDS, breast cancer, and road traffic accidents. HAI is considered as the eighth most common cause of death in the United States. The mortality rate also varies depending on the patient population and health care setting.

Common Health Care–Associated Infections

HAIs can be transmitted by the medical devices or the unclean hands of the health care workers. The infection caused by devices (such as catheter, ventilator) are catheter-related blood stream infection (CR-BSI), catheter-associated urinary tract infection (CA-UTI), ventilator-associated pneumonia (VAP), etc. Infection due to C. difficile can be transmitted among the patients due to mal hand hygiene practice of the health care worker. The various bacterial pathogens causing HAI are MRSA, VRE, vancomycin-resistant Staphylococcus aureus, M. tuberculosis, Acinetobacter spp., P. aeruginosa, B. cepacia, C. difficile, Clostridium sordellii, Escherichia coli, and Klebsiella spp. Infections due to hepatitis virus, HIV, influenza virus, and norovirus are the common infections acquired in the hospital settings.

Infections in Operating Rooms and Intensive Care Units

SOURCE

The infections in operating rooms (ORs) mainly appear as surgical site infections (SSIs). The route of entry of the pathogens can be either air born or by contact of the contaminated instruments and hands of health care provider. The majority of SSIs occur due to transfer of bacteria from air to the wound and less frequently with the contact. The bacteria from the patient get into the air and later settle down on the surface of the floor/instruments/health care provider's hand, etc. Ventilation of the OTs dramatically reduce the incidence of postoperative infections. It is also observed that the number of airborne infection increases with the number of members and movements inside the OT setup. The patients, in the ICU, have severe illness, impaired defense mechanism, and on the invasive devices (e.g., ventilators, IV catheter, urinary catheter). Extreme of age, multiple trauma, burn, and abdominal surgery are the commonest risk factors for the patients to acquire NI during their stay. Prolong hospitalization, immobilization, overcrowding, low patient–nurse ratio, and poor infection control practices enhance the infection rate in the ICUs. Common infections that occur in the ICUs are VAP, CR-BSI, CR-UTI, SSI, C. difficile–associated diarrhea, etc. Device-related infections are most common among all.

Measures to Reduce Health Care–Associated Infection

SSIs are the commonest HAI infections that occur in the OT due to faulty preventive protocols. These infections are difficult to treat and life threatening also. Hence, recommendation of guidelines and application of the methods are required to prevent SSIs in the ORs.

Design of Operating Room

Designing the ORs is a complex task and should be done as per the recommendations. At least one anesthesiologist should be in the team while designing ORs. The aim is to give maximum benefit to the patients arriving to the ORs for various diagnostic, therapeutic, and palliative procedures. The ORs should be separated from the main hospitals. The floors and walls should be smooth and covered with antistatic material, so that lesser amount of dust particles can be absorbed. Broadly, the ORs can be described in three sections. From outer to inner disposal zone, clean zone, and aseptic zone. The innermost protective zone should have maximum negative pressure and least bacteriological count in comparison to other zones. The inner negative pressure allows the free flow of germ-free air from outside to inside.

Outer/disposal zone: outer disposal area of the ORs.

Clean zone: a semirestricted area contains store room, anesthetic room, recovery room, scrub room, and rest room for the staffs. The exit of the clear zone should be through the outer zone.

Aseptic zone: it should be restricted to the working team. It includes the OT and the area for storage of sterile equipment.

Ventilation and Cleaning of the Operating Rooms

In the ORs, air flow system or ventilation removes the majority of airborne bacteria. ORs should be designed properly for adequate ventilation. The air flow system is composed of ventilation, direction of air flow, maintenance of pressure, air filtration, etc. Clean filtered air and adequate ventilation of indoor air are the two key factors in reduction of airborne infection. Ventilation helps in maintenance of the temperature and humidity. The OR's temperature should be kept 1°C lesser than the outside and maintained in between 18 and 24°C. About 40–60% of relative humidity should be maintained inside the ORs for the comfort of the staff. Ventilation causes exchange of air, and thereby removes all the bacteria released from the patient during the operative procedure and dilutes the anesthetic gases from the ORs. The direction of air flow should remain from the ORs to the main corridor. It has been observed that turbulent air flow exchanges the air more efficiently than low-velocity unidirectional flow. The anesthetic gases should be removed other than the main exhaust. Filters with an efficiency of 20–40% are adequate at the entry side. HEPA filters, which can remove particles of size >0.3 μm with efficiency of >99.97% are used as the second-line filters near the ORs, ICUs, burn wards etc. Air from the HEPA filters flow unidirectionally (vertically or horizontally) causing few airborne bacteria at the operation site. Exponential laminar air flow has been designed to overcome the limitation. The recommended air change rates (ACHs) in ORs and recovery room are 20 ACH/h and 12–15 ACH/h, respectively. In ultraclean ventilated theaters, the minimum ACH should be 37 ACH/h.

Cleaning of the OT complex should be done either with vacuum or by wet mopping. Try to keep the surfaces and floors dry, when the room is in use. The equipment inside the room should be covered during the cleaning process. All the inanimate objects (like tables, chairs, trolleys, sink door handler, etc.) should be cleaned with an EPA-approved LLD detergent. Mopping should be regularly done at the beginning of each day before the OT process is going to start. The OT tables, trolley tops, lamps, etc., should be cleaned between use for each patient using LLD.

MICROBIOLOGICAL SAMPLING

Routine bacteriological surveillance of ORs should be done for monitoring the presence of various airborne pathogens. The rate of microbial contamination in a well-designed, properly filtered, ventilated, and disinfected OT is much less. In conventionally ventilated theaters, microbiological sampling is done in the following way. Close all the doors of the ORs and keep empty for 1 h. Using an air sampler, >250 L [recommended amount is 1 m3 air (100 L)] air should be exposed to the no-selective culture medias. At least two samples should be taken per OR. The colony count should not exceed 50–150 CFU/m3. As per the UK guidelines, with a 5-min exposure, the standard bacterial count should not exceed 35 bacterial and/or fungal particles per cubic metre of ventilating air. Similarly, the Geneva guidelines takes the standard of 25 CFU/m3 for an empty OR and 180 CFU/m3 for 5-min exposure. The working condition of HEPA filters should be checked first. In properly functioning HEPA filters, the amount of airborne microbial contamination is very low. In unidirectional air flow zone, microbiological sampling is done from the four corners of the perimeter zone, one from the center and four from the inner corners of the inner zone. This zone should not contain more than 0.5 CFU/m3.

Health Care–Associated Infections in the Intensive Care Units and Its Prevention

NIs or HAIs are a major cause of morbidity and mortality in the medical setup. The patients in the ICUs are observed to be more ill, receiving multiple high antibiotics, on multiple devices, and exposed to multiple antibiotic-resistant colonizers. The incidence of HAIs is found to be five to six times higher in the ICUs than in the general wards. The overall rate of ICU infection varies from 5% to 35%

of which approximately 25% are hospital-acquired infections. The most common infections observed in the ICU setup are ventilator-associated infection (VAP), CR-BSI, CA-UTI, SSI, C. difficile-associated diarrheas, etc. Among these infections device-related infection (VAP, CA-BSI, CA-UTI) constitute 80%. Various risk factors for the development of infections are divided into host-related and hospital-related factors.

The host-related factors include severity of diseases, extreme of ages, immunocompromised conditions (malignancy/transplant/organ failure/HIV infection), burn injury, trauma, and extensive surgeries. The factors related to the hospital or treatment are device related, treatment on immunosuppressive drugs, multiple blood transfusion, hemodialysis, parenteral nutrition, prolong immobilization, etc. Since most of the infections are caused by the colonizers, the common pathogens isolated in these infections are MRSA, VRE, ESBL-producing gram-negative bacteria, Stenotrophomonas maltophilia, fluconazole-resistant Candida spp., etc.

HAIs in the ICUs can be prevented by the application of recommended preventive measures. The HAI rate has been reduced to one-third with proper applicability of infection control programs. One should have adequate knowledge about the risk factors, source of infection, type, and causation of infections to plan and establish the preventive strategies.

BUNDLE APPROACH FOR REDUCING INFECTION

Bundle approach is an evidence-based group approach of preventive measures, which is found to be more effective when executed together. Studies confirm that combined interventions result in better outcome than individual. These bundle approaches are small (three to five), straightforward practices performed in groups. These are mainly applied for the NIs such as CR-BSI, VAP, CA-UTI, and SSI. However, the rate of reduction of these infections also depends on the baseline rate of infection of that health care center, staff adherence to the tactics, and the preventive measures chosen for the bundle.

CR-BSI is an important cause of mortality and morbidity in patients admitted to the ICUs. The number increases with the increase in handling, duration of insertion, number of manipulation, number of lumens, etc. Studies have proved that a large proportion of infections are preventable by applying the various control measures. All the elements of the bundle must be executed at the same point of time. These include hand washing, full-barrier precaution during the insertion of central line catheter, cleaning of the skin with chlorhexidine solution every time, frequent infusion of heparin or heparinlike substance through total parenteral nutrition to prevent the formation of fibrin, removal of unnecessary catheters, and avoidance femoral site catheterization.

VAP is another important cause of ICU-acquired infection. It not only causes mortality but also is responsible for substantial cost of treatment. Various recommendations for the prevention of VAP include (1) appropriate cleaning, disinfection, and sterilization of ventilator equipment, (2) maintenance of ventilator circuits, and (3) routine care of patients requiring ventilation.14 The cleaning, disinfection, and sterilization of the equipment have been discussed earlier in the chapter. The preventive measures taken for the patients to reduce the rate of VAP are use of orotracheal intubation, noninvasive ventilation, minimization of the duration of ventilation. The patient should be maintained in semirecumbent position (30–45 degrees elevation of head of the bed). Regular oral care should be performed. CA-UTI is another common cause of infection in the ICUs. Staff must be trained enough to differentiate asymptomatic bacteriuria from infection due to catheterization. Regular education regarding repeated hand washing before and after catheterization, maintenance of a closed system, maintaining free urine flow, and securing the catheter position correctly should be given to the health care workers for the prevention of these infections. SSI is a very common and life-threatening infection during ICU stays. There are various evidence-based guidelines recommended by the CDC. Preoperative patient preparation, management of the colonizers, correct surgical technique, adequate antimicrobial prophylaxis, adequate sterilization and disinfection of the environmental surfaces, proper aseptic techniques, as well as washing surgical hands before and after the operative procedure are various preventive measures that can be taken to reduce the incidence of SSIs.

About one-third of the infections in the ICUs can be prevented by implementation of recommended preventive measures. Apart from education of the health care personnel and standard precaution measures, other precautions like contact precaution, airborne precaution, and droplet precaution should be taken to prevent various infections. Antibiotic-associated diarrhea due to C. difficile, and infection due to E. coli (O157:H7), Shigella spp., hepatitis A, rotavirus, etc., can be transmitted by contact with infected patients or items. Similarly infections transmitted by droplet such as tuberculosis, varicella, measles, viral hemorrhagic fever, and influenza can be prevented by taking adequate precaution. Use of antibiotics should be restricted in the ICU setup. Minimum precautions like heating the water before use, regular cleaning, and maintenance of water tanks can be done to prevent water-borne infections.

Difference Between Disinfection and Sterilization

Disinfection and sterilization are the two common methods of killing or inhibiting the growth of microorganisms. Microbes can cause a number of diseases in humans and animals. They are also responsible for food spoilage. Therefore, it is beneficial to kill or inhibit their growth to minimize their harmful effects. This is achieved by sterilization or disinfection. However, both these processes are based on different principles.

Let us have a detailed look at the difference between disinfection and sterilization.

Disinfection vs Sterilization

Following are the important difference between disinfection and sterilization:

Disinfection	Sterilization
In this, the number of harmful microbes is minimized to a negligible level.	In this, the medium is made completely free from all microbes.
It kills only vegetative cells and not the spores.	It kills both vegetative cells and spores.
Wounds are disinfected – with agents such as hydrogen peroxide or rubbing alcohol.	Wounds cannot be sterilized – as it may kill surrounding healthy cells.
Disinfection only reduces the effect of microbes.	Sterilization completely rids microbes from the surface
Chemical methods are used for disinfection	Combination of heat, irradiation, high pressure, chemical and physical methods are used for sterilization
Phenol, alcohol, chlorine, iodine are some of the disinfecting agents.	High temperatures, steam, radiation, filtration are some of the sterilization techniques.
Only adequate cleanliness	Extreme cleanliness
Used in daily life.	Used primarily for medical and research purposes.
For eg., pasteurization, disinfecting urinals, etc.	For eg., sterilization of instruments used during surgery by autoclaving.

What is Sterilization?

Sterilization is a process mainly used to kill all forms of microorganisms and its spores. This process is carried out to maintain a sterile environment. It is usually done through combinations of heat, irradiation, filtration, high pressure etc.

WHAT IS DISINFECTION?

Disinfection is a chemical process used to eliminate or inhibit the growth of bacteria and other pathogens that may cause infectious diseases in humans and animals. Disinfectants are the chemical solutions used to make substance infection-free. These include alcohol, iodine, chlorine, etc.

Similarities Between Disinfection and Sterilization

Disinfection and sterilization have a few things in common. These include:

- Both are used to decontaminate objects and surfaces
- Both processes can kill (most) harmful microbes.

CONCLUSION

The main takeaway here is that the aim of disinfection and sterilization are different. Disinfection aims to kill or inactivate harmful microorganisms, but leaves some microbes intact. Sterilization, on the other hand, aims at killing all the microbes. And each form of decontamination has its functional use, as stated in the comparison chart.

5

HIP REPLACEMENT

Hip replacement is a surgical procedure performed in dogs and cats as a salvage procedure, to alleviate severe pain in the hip due to, for example, hip dysplasia or irreparable bone fracture. The procedure replaces the head of the femur and the acetabulum with prosthetic implants. Because animals under about 40 pounds (18 kg) carry their own weight with little strain on each leg, hip modification surgeries are often sufficient to restore hip function in many cases. As a result, while hip replacement on animals can be seen in any animal of any size, from cats upwards, it is most often performed in the medium-large breeds of dogs.

An animal suffering from hip dysplasia may have been in some degree of chronic pain all its life from a very young age but may show no overt sign of pain. Rather, they have adapted to the pain over an extended period, and learned to live with it. As a result, in many animals, successful hip replacement gives animals a level of playfulness and happiness not previously seen.

As with all surgeries, results vary, and there is both risk and uncertainty involved.

IMPLANT MATERIAL

Animal hip replacements are usually made from the same material as human replacements were, historically, made—a metal (cobalt chrome or titanium) femoral component, and an ultra high molecular weight polyethylene acetabular component.

Since animals have a shorter life than humans, their hip replacements can safely be assumed to be lifelong and will not wear out or need replacing. By contrast, human hip replacements are sometimes deferred until older, to avoid the possible need for replacement later on in life.

Considerations And Aftercare

Other treatments may give a good quality of life with fewer attendant risks. Surgery for dogs, as with humans, is a final alternative when non-invasive methods are unable to treat a condition, since they are irreversible and carry risk. Conditions such as hip dysplasia and arthritis can often be treated with appropriate medications that help the body handle pain, inflammation, or joint wear and tear.

Animals which are prospective candidates for hip replacement should have no other spinal, neural, or rear leg abnormalities, in order to be significantly helped by a hip joint procedure.

Since the reduced joint mobility seen in conditions such as dysplasia may result in loss of muscle mass and quality as a dog ages, there is often an advantage in having hip replacement whilst the dog is at an early age, while muscle is more likely to re-develop, rather than in old age when convalescence is longer and more difficult. However, this is a major surgery taking several months to fully recuperate, involving the large muscle groups of the hips, and is irreversible. Whilst it has a high success rate (circa 95%) in the hands of a good surgical team, even in older dogs, it is therefore often recommended to avoid it until quality of life is seriously affected beyond the capability of medication to control.

Thus the benefits of hip replacement at a younger age must be set against the risks and the existing quality of life attainable with medication, lifestyle change, or non-surgical handling. Usually a course of medications is tried in any event, to assess how the animal responds to them.

An animal will usually need a minimum of 2 months convalescence to recover from hip replacement surgery. This allows the new artificial acetabular cup (the caput, or hip socket) to bond properly to the pelvis (hip bone). During this period, the animal must be restricted to carefully limited mobility and exercise, as the joint is still bonding and new bone is being laid down. So animals must be prevented from over exercise, or from climbing, jumping or putting any strain whatsoever beyond gentle use, on the joint. The animal must also be kept away from slippery or smooth flooring such as tiles, marble or polished wood since these put considerable lateral (sideways) strain on the hips. This can be a difficult process for a dog as many usual behaviors must be strictly prevented for this period. A pet crate may be a sensible precaution, if in doubt.

Usually, in the case of a double replacement (both hips), whenever possible one hip is operated on at a time and allowed to heal before the other is replaced. This ensures that there is always the maximum natural support during the healing process, although bilateral (double) hip replacements are possible and can be performed if appropriate.

Variations in vets' views

Some vets will recommend hip replacement at any age over puberty, if suitable conditions are met. Others view it as a surgery to be avoided at all costs unless there is no alternative. The issue here appears to be threefold: consideration of cost, avoidance of surgery where possible, and historical reasons that this has usually in the past been a treatment afforded older dogs and not usually considered for young and middle aged animals or less advanced conditions.

Aftercare recommendations vary also. Some vets recommend as little as 1–2 days in hospital after replacement, others as much as 4–7 days. All vets agree that movement must be restricted for a significant time after surgery. Some suggest that normal activity can be resumed after 5 weeks, others state it is unsafe until after 8 weeks. In both cases, a cautious approach will probably do no harm, if there is doubt.

HIP DYSPLASIA (CANINE)

In dogs, hip dysplasia is an abnormal formation of the hip socket that, in its more severe form, can eventually cause lameness and arthritis of the joints. It is a genetic (polygenic) trait that is affected by environmental factors. It is common in many dog breeds, particularly the larger breeds, and is the most common single cause of arthritis of the hips.

During growth, both the ball (the head of the femur, or thighbone) and the socket in the pelvis (acetabulum) must grow at equal rates. In hip dysplasia, this uniform growth during puppyhood does not occur. The result is laxity (looseness) of the joint, followed by degenerative joint disease (DJD) or osteoarthritis (OA), which is the body's attempt to stabilize the loose hip joint.

Normal hip anatomy

In the normal anatomy of the hip joint, the almost spherical end of the femur head (the caput, or caput ossis femoris) fits into the acetabulum (a concave socket located in the pelvis). The bony surfaces of the femur head and of the acetabulum are covered by cartilage. While bones provide the strength necessary to support body weight, cartilage ensures a smooth fit and a wide range of motion. Normal hip function can be affected by congenital conditions such as dysplasia, trauma, and by acquired diseases such as osteoarthritis and rheumatoid arthritis.

Dysplastic hip anatomy

The hip could have major contractions from dysplasias. The caput is not deeply and tightly held by the acetabulum. Instead of being a snug fit, it is a loose fit, or a partial fit. Secondly, the caput or acetabulum are not smooth and round, but are misshapen, causing abnormal wear and tear or friction within the joint as it moves. The body reacts to this in several ways. First, the joint is continually repairing itself and laying down new cartilage. However, cartilage repair is a relatively slow process, the tissue being avascular, so the joint may suffer degradation due to the abnormal wear and tear, or may not support the body weight as intended. The joint becomes inflamed and a cycle of cartilage damage, inflammation and pain commences. This is a self-fueling process, in that the more the joint becomes damaged, the more damage it will cause to the surrounding tissues and bones. The inflammation also causes further damage. The bones of the joint may also develop osteoarthritis, visible on a radiograph as small outcrops of bone, which further degrade the joint. Osteoarthritis is a degenerative disease marked by the breakdown of cartilage between joints resulting in painful bone-to-bone contact.

The underlying deformity of the joint may get worse over time, or may remain static. A dog may have good radiographs and yet be in pain, or may have very poor radiographs and have no apparent pain issues. The hip condition is only one factor to determine the extent to which dysplasia is causing pain or affecting the quality of life. In mild to moderate dysplasia it is often the secondary effects of abnormal wear and tear or arthritis, rather than dysplasia itself, which is the direct causes of visible problems.

Causes and effects

Hip dysplasia may be caused by a femur that does not fit correctly into the pelvic socket, or poorly developed muscles in the pelvic area. Large and giant breeds are most susceptible to hip dysplasia (possibly due to the body mass index (BMI) of the individual animal), though many other breeds can suffer from it. The Orthopedic Foundation for Animals maintains a list of top 100 breeds affected.

To reduce pain, the animal will typically reduce its movement of that hip. This may be visible as "bunny hopping", where both legs move together, or less dynamic movement (running, jumping), or stiffness. Since the hip cannot move fully, the body compensates by adapting its use of the spine, often causing spinal, stifle (a dog's knee joint), or soft tissue problems to arise.

The causes of hip dysplasia are considered heritable, but new research conclusively suggests that environment also plays a role. To what degree the causality is genetic and what portion environmental is a topic of current debate. Neutering a dog, especially before the dog has reached an age of full developmental maturity, has been shown to almost double the chance he or she will develop hip dysplasia versus intact dogs or dogs that were neutered after reaching adulthood. Other environmental influences include overweight condition, injury at a young age, overexertion on the hip joint at a young age, ligament tear at a young age, or repetitive motion on forming joint (e.g. jogging with puppy under the age of 1 year). As current studies progress, greater information may help provide procedures to effectively reduce the occurrence of this condition.

It is most common in medium-large pure bred dogs, such as Newfoundlands, German Shepherd Dogs, retrievers (such as Labradors, Tollers, or Goldens), Rottweilers and Mastiffs, but also occurs in some smaller breeds such as spaniels and pugs.

Clinical detection and testing

Symptoms

Atrophy of thigh muscle after a two-year evolution of hip dysplasia

Dogs with hip dysplasia may exhibit the following signs and symptoms:

- Decreased activity
- Decreased range of motion
- Difficulty or reluctance rising, jumping, running, or climbing stairs
- Lameness in the hind end
- Looseness in the joint
- Narrow stance
- Swaying, "bunny hopping" gait
- Grating in the joint during movement
- Loss of thigh muscle mass
- Noticeable enlargement of the shoulder muscles as they compensate for the hind end
- Pain
- Stiffness or soreness after rising from rest
- Subluxation or dislocation of the hip joint

Affected dogs can show clinical signs as early as seven months of age, but most do not until one to two years of age. In part this is because the underlying hip problem may be mild or severe, worsening or stable, and the body may be more or less able to keep the joint in repair well enough to cope.

Diagnosis

Hip dysplasia is diagnosed with radiographs of the pelvis. There are several standardized systems for categorising dysplasia, set out by reputable bodies. The most widely used systems include the following:

Orthopedic Foundation for Animals (OFA)

Radiographs can be sent to OFA for grading and certification. This system rates a dog's hip joint on a seven-point scoring system. The test relies on interpretation of a radiograph of the dog's hips, which are then assigned a score by three independent radiologists: Excellent, Good, Fair, Borderline, Mild, Moderate and Severe.

PENNHIP

PennHIP requires the dog to be anesthetized. Three radiographs are taken to measure the hip joint laxity. A score between 0-1 is assigned, with 0 being very tight hips and 1 being very loose. The score is based on a measurement of the hip's distraction index (DI). The distraction index is an indication of the "percent out of joint" that the femoral head is displaced from the acetabulum. For example, a DI of 0.15 means that the head of the femur is 15% out of joint.

British Veterinary Association (BVA)

Each hip joint is assessed by a BVA panel of experts that assign points based on nine aspects of the joint. The degree to which a dog is affected by hip dysplasia is represented by a score given to each hip. Scores for each hip are added together to get an overall hip score. Score ranges from zero to 106 (zero to 53 for each hip), with a score of zero representing the least degree of hip dysplasia and 53 representing the most.

It is also common to X-ray the spine and legs, as well as the hips, since soft tissues can be affected by the extra strain of a dysplastic hip, or there may be other undetected factors such as neurological issues (e.g. nerve damage) involved.

Other conditions with similar symptoms

The following conditions can give symptoms very similar to hip dysplasia, and should be ruled out during diagnosis:

- Cauda equina syndrome (i.e. lower back problems)
- Cranial (anterior) Cruciate ligament tears

- Other rear limb arthritic conditions
- Osteochondritis dissecans and elbow dysplasia in the forelimbs are difficult to diagnose as the animal may only exhibit an unusual gait, and may be masked by, or misdiagnosed as, hip dysplasia.

A dog may misuse its rear legs, or adapt its gait, to compensate for pain in the forelimbs, notably osteoarthritis, osteochondritis (OCD) or shoulder or elbow dysplasia, as well as pain in the hocks and stifles or spinal issues. It is important to rule out other joint and bodily issues before concluding that only hip dysplasia is present. Even if some hip dysplasia is present, it is possible for other conditions to co-exist or be masked by it.

Treatment

There is no complete cure, although there are many options to alleviate the clinical signs. The aim of treatment is to enhance quality of life.

Most dogs with hip dysplasia do not need surgery as they do very well with a home physical therapy program, keeping trim and strong, and using pain medication as needed.

If the problem cannot be controlled with medications, then often surgery is considered. There are traditionally two types of surgery - those which reshape the joint to reduce pain or help movement, and hip replacement which completely replaces the damaged hip with an artificial joint, similar to human hip replacements.

Non-surgical interventions

Non surgical intervention is dependent on many factors including age, weight, degree of hip laxity, lifestyle of the owner and their tolerance for the cost incurred for medication and physical therapy.

Maintenance of minimal body weight

Weight control is often "the single most important thing that we can do to help a dog with arthritis," and "reducing the dog's weight is enough to control all of the symptoms of arthritis in many dogs." With weight control, the goal is to prevent the dog from becoming overweight to reduce mechanical stresses applied to the hip joints. In general terms, the ribs should be easily palpated and there should be an indentation in front of the hip bones.

Exercise routine

Reasonable exercise stimulates cartilage growth and reduces degeneration, and also regular walks taken in the early stages of dysplasia can help prevent loss of muscle mass to the hips. Exercise should fit an individual dog's maximum intensity level with the goal to maintain muscle tone and cardiovascular function without causing pain, stiffness, and inflammation to the joint. Exercise also improves joint range of motion which in turn keeps the dog more comfortable. Swimming, because it is a non-weight bearing exercise, can be a very useful means of maintaining muscle tone and range of motion without placing concussive forces on the joint, though do know that swimming is primarily a front-legged exercise.

Medication

Medication can reduce pain and discomfort, and also reduce damaging inflammation. Most often, medication consists of a non-steroidal anti-inflammatory drug (NSAID), which doubles as an anti-inflammatory and painkiller. Typical NSAIDs used for hip dysplasia include carprofen and meloxicam (often sold as Rimadyl and Metacam respectively). Both are used to treat arthritis resulting from dysplasia, although other NSAIDs such as tepoxalin (Zubrin) and prednoleucotropin ("PLT", a combination of cinchophen and prednisolone) are sometimes prescribed.

Diet supplementation

Incorporating omega-3 fatty acids such as docosahexaenoic acid (DHA) and eicosapentaenoic acid (EPA) into the diet can result in improved symptoms of the disease. Omega-3 fatty acids can help decrease inflammation that occurs from osteoarthritis, as well as improve the locomotion of dogs who have the disease. EPA and DHA can be supplemented into the diet through fish oils.

Glucosamine and chondroitin sulfate are nutraceuticals that can also be added into the diet to help treat osteoarthritis caused by hip dysplasia. Both nutraceuticals help with improvement of cartilage, joint health and repairing of tissues. A glucosamine-based nutritional supplement may give the body additional raw materials used in joint repair. Glucosamine can take 3–4 weeks to start showing its effects, so the trial period for medication is usually at least 3–5 weeks. In vitro, glucosamine has been shown to have negative effects on cartilage cells.

Another nutrient that can help improve the structural support of the body is vitamin C. Vitamin C contributes to the building blocks of collagen that can help to strengthen the joints.

Other interventions

Canine massage may alleviate discomfort, decrease muscle soreness, and spasms, and help move lymph and nutrients through the system.

Some attempts have been made to treat the pain caused by arthritic changes through the use of laser therapy, in particular class IV laser therapy. Well-controlled clinical trials are unfortunately lacking, and much of the evidence for these procedures remains anecdotal.

The use of pressure-reducing pet beds, ramps, stairs, and steps built with wood, plastic, metal, or foam prevent further damage. Hip hammocks have been tested and proven effective in aiding dogs suffering from hip dysplasia.

Surgical interventions

Surgical procedures attempt to modify, repair, or replace the hip joint in order to allow pain-free usage.

- Femoral head ostectomy (FHO) is when the head of the femur is removed but not replaced. Instead, the resulting scar tissue from the operation takes the place of the hip joint. This false joint is less stable with a reduced range of motion than a normal joint, causing an abnormal gait. Nevertheless, pain relief with adequate function can be achieved. The procedure can be performed in all dogs of all sizes, but there are usually better long-term success rates in smaller dogs less than 20 kg (44 lbs). This procedure is used when there is significant osteoarthritis and a total hip replacement is cost-prohibitive.

- Triple pelvic osteotomy (TPO) involves bone cuts in the pelvis so that the socket portion of the joint can be rotated over the ball; the bones are then stabilized with a bone plate. This procedure is indicated for dogs under 10 months of age. A TPO can be performed to reestablish joint stability and encourage normal joint development. This procedure is not indicated if osteoarthritis is already present. Recovery time is about six weeks.

- Hip replacement has the highest rate of success, especially in severe cases, since it completely replaces the faulty joint. It usually restores complete mobility if no other joint is affected, and also completely prevents recurrence. It is ideal for dogs over 10 months old that already have established osteoarthritis and can no longer be medically managed. Hip replacement is also the preferred clinical option for serious dysplasia in animals that weigh 40–60 lb (18–27 kg).

- Juvenile pubic symphysiodesis (JPS) is a procedure for very young dogs that manipulates the way the pelvis grows to create a tighter hip. It involves cauterizing the growth plates of the pelvis to halt vertical growth. To compensate, the rest of the pelvis grows outward, in a manner which enhances the "socket" of the hip and provides better support than the dog would have had naturally. Since it relies on growth in puppyhood, it has a very tight window for surgery – between 4 and 5 months old. This is compatible with hip scoring of puppies at 4 months.
- Capsular neurectomy is a procedure in which the hip joint capsule is de-nerved to reduce pain in the hip. This allows the dog to exercise moderately with less pain, thus preventing the leg muscles from weakening from disuse and providing less support to the bad joint. Both hips can be done in one surgery. This surgery should not prevent a future hip replacement if a more complete fix is desired..

OSTEOTOMY

An osteotomy is a surgical operation whereby a bone is cut to shorten or lengthen it or to change its alignment. It is sometimes performed to correct a hallux valgus, or to straighten a bone that has healed crookedly following a fracture. It is also used to correct a coxa vara, genu valgum, and genu varum. The operation is done under a general anaesthetic.

Osteotomy is one method to relieve pain of arthritis, especially of the hip and knee. It is being replaced by joint replacement in the older patient.

Due to the serious nature of this procedure, recovery may be extensive. Careful consultation with a physician is important in order to ensure proper planning during a recovery phase. Tools exist to assist recovering patients who may have non weight bearing requirements and include bedpans, dressing sticks, long-handled shoe-horns, grabbers/reachers and specialized walkers and wheelchairs.

Osteotomies of the Hip

Two main types of osteotomies are used in the correction of hip dysplasias and deformities to improve alignment/interaction of acetabulum – (socket) – and femoral head (femur head) – (ball), innominate osteotomies and femoral osteotomies. The bones are cut, reshaped or partially removed to realign the load-bearing surfaces of the joint.

a. X-ray of a hip after capsulorrhaphy and Dega surgery.

b. Same hip after addition of Chiari medial displacementosteotomy.

Adjustments are made to part of the hip-bone. Many operating methods and variations have been developed. They are defined by the type of cut and adjustment made. Some acetabular procedures are named after the surgeons who first described them as Salter (R. Salter), Dega (W. Dega), Sutherland (D.H. Sutherland), Chiari (K. Chiari): other names one may encounter are Ludlov, P. Pemberton, and James B. Steele. Some are named after the shape of cut (e.g. Chevron, Wedge) or the way the bones are aligned (Dial=old style rotary dial phone).

A femoral derotation osteotomy can be performed to correct version abnormalities such as excessive anteversion or retroversion of the hip joint. Excessive anteversion of the femur results in anterior instability of the hip joint while excessive retroversion results in femoroacetabular hip impingement.

A subtrochanteric blade plate or an intramedullary rod can be used to stabilize the osteotomy site in a femoral derotation osteotomy until compete bone healing is achieved; an approach employing an intramedullary rod is much less invasive than one using a subtrochanteric blade plate.

Femoral osteotomies, as the name indicates, involves adjustments made to the femur head and/or the femur.

Osteotomy of the knee

Knee osteotomy is commonly used to realign arthritic damage on one side of the knee. The goal is to shift the patient's body weight off the damaged area to the other side of the knee, where the cartilage is still healthy. Surgeons remove a wedge of the tibia from underneath the healthy side of the knee, which allows the tibia and femur to bend away from the damaged cartilage.

A model for this is the hinges on a door. When the door is shut, the hinges are flush against the wall. As the door swings open, one side of the door remains pressed against the wall as space opens up on the other side. Removing just a small wedge of bone can "swing" the knee open, pressing the healthy tissue together as space opens up between the femur and tibia on the damaged side so that the arthritic surfaces do not rub against each other.

Osteotomy is also used as an alternative treatment to total knee replacement in younger and active patients. Because prosthetic knees may wear out over time, an osteotomy procedure can enable younger, active osteoarthritis patients to continue using the healthy portion of their knee. The procedure can delay the need for a total knee replacement for up to ten years.

SURGERY

The location of the removed wedge of bone depends on where osteoarthritis has damaged the knee cartilage. The most common type of osteotomy performed on arthritic knees is a high tibial osteotomy, which addresses cartilage damage on the inside (medial) portion of the knee. The procedure usually takes 60 to 90 minutes to perform.

During a high tibial osteotomy, surgeons remove a wedge of bone from the outside of the knee, which causes the leg to bend slightly inward. This resembles the realigning of a bowlegged knee to a knock-kneed position. The patient's weight is transferred to the outside (lateral) portion of the knee, where the cartilage is still healthy.

After regional or general anesthesia is administered, the surgical team sterilizes the leg with antibacterial solution. Surgeons map out the exact size of the bone wedge they will remove, using an X-ray, CT scan, or 3D computer modeling. A four- to five-inch incision is made down the front and outside of the knee, starting below the kneecap and extending below the top of the shinbone.

Guide wires are drilled into the top of the shinbone (tibia plateau) from the outside (lateral side) of the knee. The wires usually outline a triangle form in the shinbone.

A standard oscillating saw is run along the guide wires, removing most of the bone wedge from underneath the outside of the knee, below the healthy cartilage. The cartilage surface on the top of the outside (lateral side) of the shinbone is left intact. The top of the shinbone is then lowered on the outside and attached with surgical staples or screws, depending on the size of the wedge that was removed. The layers of tissue in the knee are stitched together, usually with absorbable sutures.

Rehabilitation and prevention

A fall or torque to the leg during the first two months after surgery may jeopardize healing. Patients must exercise extreme caution during all activities, including walking, until healing is complete.

After rehabilitation, preventing osteoarthritis involves slowing the progression and spread of the disease. Maintaining aerobic cardiovascular fitness has been an effective method for preventing the progression of osteoarthritis. Light, daily exercise is much better for an arthritic knee than occasional, heavy exercise.

It is especially important to avoid any serious knee injuries, such as torn ligaments or fractured bones, because arthritis can complicate knee injury treatment. High-impact or repetitive stress sports, like football and distance running, should be avoided.

Because osteoarthritis has multiple causes and may be related to genetic factors, no universal prevention tactic exists.

General recommendations include:

- Keeping a slight bend in the knees will take the pressure off during standing.
- Avoid activities that causes pain which lasts over an hour.
- Perform controlled range of motion activities that do not overload the joint.
- Avoid heavy impact on the knees during everyday and athletic activities.
- Gently strengthen thigh and lower leg muscles to help protect the bones and cartilage in the knee.
- Non-contact activities keep joints and bones healthy and maintain fitness over time. Exercise also helps promote weight loss, which can take stress off knees.

OSTEOTOMY OF THE JAW

This is performed to realign the mandible (lower jaw) or maxilla (upper jaw) with the rest of the skull and/or teeth. This is usually performed to correct skeletal malocclusions, that is discrepancies in tooth position that cannot be corrected by simple orthodontic movement, and realignment of the temporomandibular joints, or to correct facial deformities such as mandibular retrognathia. There is little scarring, and all of the surgery takes places inside of the mouth. Orthodontic braces may have to be worn pre- and post- operation to realign the teeth to match the newly realigned jaw.

Osteotomy of the chin

Chin osteotomy is most often done to correct a vertically short chin. As opposed to putting an implant on top of the chin bone to bring it forward, an alternative approach is to cut the chin bone itself and bring it forward or other directions as well. It can also be used to lengthen the chin (which is more difficult with an implant) or to shorten or narrow a chin. (which is impossible with an implant).

Chin osteotomies (cutting the bone and moving it) are done through an incision inside the mouth. It is technically more difficult than an implant and has more swelling and recovery than a simple chin implant. Also, there is usually temporary loss of feeling of the lip and chin after that takes several weeks to months for full return of sensation.

Veterinary osteotomy procedures

In veterinary medicine, osteotomies are frequently performed to address rupture of the canine cranial cruciate ligament, which is analogous to the anterior cruciate ligament. The tibial plateau leveling osteotomy and tibial tuberosity advancement are two of the most common osteotomy procedures performed in the United States. Recovery is often 6–8 weeks and the osteotomy can be filled with autologous bone grafts, scaffolds (hydroxyapatite, TR Matrix, etc.) or ceramics.

6

SURGICAL ONCOLOGY

Surgery is often the preferable option for local control of cancer. Successful removal of a localized cancer can be one of the best approaches to treat cancer patients and provide them with improved quality of life.

There are several reasons to choose a surgical approach.

- **Diagnostic Intent:** The first is to gather tissue for diagnosis. In this case, surgery is used to remove tissue for biopsy. The tissue is examined by diagnostic pathologists such that the most accurate diagnosis and treatment protocol can be made.
- **Curative Intent:** The second reason is to use surgery for curative intent. In this instance, the surgeon's intent is to remove cancerous tissue or tumors with wide margins, with no cancer left behind. This approach can include other treatment modalities such as radiation and chemotherapy.
- **Cytoreductive Surgery:** The third strategy is cytoreductive surgery. This involves planned incomplete excision of a tumor and is indicated when complete excision carries unacceptable consequences. The goal is to is to "downstage" the disease to microscopic level, i.e. decrease tumor burden, and therefore enhance the response to an adjuvant therapy (e.g. radiation therapy or chemotherapy).
- **Palliative Care:** The fourth surgical strategy is for palliative care. In this instance, the purpose of the surgery is to improve the quality of the patient's life. While this surgical approach may improve the quality of the patient's life, it may not necessarily extend the life of the patient.

Cancer is one of the major causes of morbidity in veterinary patients, and surgical removal of tumors one of the most common surgical procedures performed by veterinarians. This article covers tumor evaluation and biopsy, grading and staging, and surgical approach.

Cancer is one of the major causes of morbidity in veterinary patients, and surgical removal of tumors is one of the most common surgical procedures performed by veterinarians.

In human surgical oncology, 60% of human patients battling cancer are cured by surgery alone.Likewise, in veterinary medicine, surgery is considered the most important component of treatment in dogs and cats suffering from solid tumors because, in many cases, surgery offers the best chance for improving patients' quality of life.

To help assure a positive outcome, the veterinarian needs a thorough understanding of the basic principles of surgical oncology, including:

- Characteristics of specific tumor being treated
- Surgical options available
- Potential need for adjunctive treatments, such as chemotherapy or radiation therapy.

Regardless of the treatment options sought, the overall goal of the therapeutic plan is to improve the patient's quality of life.

PATIENT EVALUATION

The patient's treatment plan should be based on:

- Complete history
- Thorough physical examination
- Diagnostic imaging
- Clinical pathology with cytology
- Histopathology, when appropriate.

Tumor Identification

The annual physical examination plays a key role in detecting neoplasia, either directly via palpation or through paraneoplastic syndromes, which may be found on routine blood and urine analysis. The physical examination allows detection of possible tumor presence and identification of concurrent conditions that may influence the treatment plan.

When a tumor is identified, record its size, location, and gross appearance, and palpate regional lymph nodes. If the lymph nodes are enlarged, firm, or immovable, the suspicion of a metastatic neoplasm rises; however, normal lymph node palpation does not guarantee lack of metastasis. Therefore, some form of lymph node biopsy is recommended in most cases.

Tumor Evaluation

Histopathologic analysis of tissues obtained by biopsy allows definitive diagnosis of:

- Tumor type (hyperplasia, metaplasia, or neoplasia)
- Biological behavior (malignant or benign)
- Aggressiveness
- Tumor grade.

There are 4 main types of tumor biopsy techniques:

1. Fine-needle aspiration (FNA) is a cost-effective, minimally-invasive procedure that is useful for differentiating non-neoplastic from neoplastic diseases and, in many situations, obtaining a definitive diagnosis.

- FNA is usually reserved for cutaneous and subcutaneous masses; some tumors, such as mast cell tumors and lymphoma, are readily identifiable on most needle aspirates.
- Intracavitary structures (especially splenic and hepatic masses) may be sampled with ultrasonography or computed tomography guidance.4-6
- However, FNA is generally the least diagnostic biopsy technique due to the minimal amount of tissue sampled and lack of organized tissue architecture, which prevents tumor grading.
- Certain tumor types, such as sarcomas, may not exfoliate cells well, leading to false-negative results. Interpret negative aspirates or aspirates with questionable results with caution, and utilize more aggressive biopsy techniques when clinically appropriate.

2. Needle-Core biopsy is more invasive than FNA but has proven to be highly diagnostic, with an accuracy of 100% for epithelial cell tumors and 94% for mesenchymal cell tumors.

- Needle-core biopsy includes 2 main techniques:
  - Tru-cut needle-core biopsy (soft tissue)
  - Jamshidi needle-core biopsy (ossified tissue).

- Typically, sedation and local anesthesia provide sufficient analgesia for biopsy collection. Caution should be used to avoid infiltrating anesthetic agents into the tumor to prevent distortion of the collected tissues.
- Collecting multiple tissue samples increases the accuracy of biopsies.

3. Incisional biopsy is utilized when less invasive techniques fail to yield a diagnosis.

- Wedge and punch biopsies are examples of incisional biopsy techniques; they are:
  - Especially useful for diagnosis of soft, friable, inflamed, and/or necrotic tumors.
  - Commonly used to sample peripheral lymph nodes and masses located on the extremities.
- For most cases, include the border between normal healthy tissue and abnormal tissue, which allows the pathologist to determine extent of tumor invasion into normal tissue.

4. Excisional biopsy allows the identified tumor to be removed in its entirety, with or without additional surgical margins.

- While this is a common approach for mass removal in veterinary medicine, it remains controversial:
  - It may provide a diagnosis and, therefore, a therapeutic plan, but the mass is removed without knowing the tumor type.
  - It is also often impossible to excise the mass without contaminating clean tissue planes or complicating future surgical procedures.
- A less invasive biopsy procedure is generally recommended to guide the surgeon in choosing the appropriate method of surgical resection.
- Excisional biopsy is best reserved for cases in which adequate surgical margins are easily achievable with biopsy alone (eg, small cutaneous mass on trunk of a dog).

Biopsy Procedure

Carefully plan the biopsy procedure to prevent spread of neoplastic cells to unaffected tissues and allow removal of biopsy tracks or scars during surgical excision of the tumor. When obtaining a biopsy from a vascular organ, such as the liver or spleen, coagulation testing should be considered.

Although a presurgical biopsy is not always required, it is essential when the:

- Surgical plan will be affected by tumor characteristics
- Tumor location will make closure of the surgical site difficult without reconstructive surgery (ie, skin flaps, grafts)
- Owner's willingness to treat may be altered based on diagnosis.

Once the biopsy is collected and properly preserved, have it processed and evaluated by a histopathologist.

GRADING & STAGING

Tumor Grading

Tumor grade is the histopathologic assessment of the tumor's aggressiveness, and includes information, such as the mitotic index, invasiveness into surrounding tissues, and degree of differentiation. This information is valuable when the surgeon is developing a treatment plan for the patient.

For some tumor types, the grading scheme may also include features, such as percent necrosis; for these neoplasms, a grade may not be accurately assessed by incisional biopsy.

Tumor Staging

If the tumor is suspected to be malignant, tumor staging establishes to what extent the tumor has invaded local, regional, or distant (systemic) tissues.

Method of tumor metastasis is highly dependent upon the tumor type: Carcinomas and round cell tumors tend to spread via lymphatic routes, while sarcomas tend to spread via hematogenous routes. Common areas where metastatic disease is identified include the lungs and lymph nodes.

In most cases, the following diagnostics are recommended:

- Three-view thoracic radiographs
- Regional lymph node palpation
- Biopsy (as mentioned earlier, lymph node palpation does not guarantee lack of metastasis).

Additional diagnostics, such as abdominal ultrasound, may be needed depending on the biologic behavior of the tumor. Other modalities, such as computed tomography or magnetic resonance, may help increase the sensitivity of detecting metastatic disease and more accurately determine the local extent of the tumor in an effort to determine prognosis and help guide resection.

SURGICAL INTERVENTION

There are 4 levels of surgical aggressiveness, or surgical "dose," to classify the extent of surgical resection (Figure 1).8,9 The most common surgical mistake is use of a surgical dose that is too low, which results from concerns about the ability to close the resultant defect.However, if the diagnosis of malignancy has been confirmed, it may be better to manage an open wound than leave tumor cells behind.

Surgical Planning

Careful surgical planning should be performed, as the first surgery is often the surgeon's best chance for achieving a positive outcome.

Fig. 6.1 A representation of a primary soft tissue tumor (blue), tumor pseudocapsule, and surrounding reactive zone (yellow), along with two satellite tumors (light blue); the lateral margins of the 4 levels of surgical resection are depicted (dotted lines): intralesional (inner most black dotted line), marginal (black dotted line at the level of the pseudocapsule), wide (white), and radical excision (chartreuse). Illustration by Michael Orencole.

Multiple factors should be considered when determining aggressiveness of surgery:

- Type of tumor
- Short- and long-term prognosis
- Ability to achieve desired surgical margins, with an acceptable level of morbidity
- Concurrent disease
- Owner wishes
- Availability of adjunctive therapy, if needed following surgery.

Surgical Doses

1. Intralesional, or debulking, surgery is incomplete resection of a tumor (inside the pseudocapsule) with residual gross disease10; it is rarely an acceptable treatment for neoplastic diseases, whether benign or malignant. This technique leaves a tumor cell burden that is often too large to result in a positive outcome even with adjunctive treatments, such as radiation therapy and chemotherapy.
2. Marginal resection is defined as removing the majority of the tumor just outside the tumor's pseudocapsule. This approach works well with lipomas and other benign tumors; however, it tends to fail with malignant neoplasms because satellite tumors are often left behind. In some cases, marginal resection can be combined with adjuvant therapies, such as radiation or chemotherapy, to provide an optimum outcome. Excisional biopsy of a malignant tumor, without prior knowledge of tumor type, is essentially an unplanned marginal resection and should be avoided in most cases. Marginal resection of malignant tumors can complicate subsequent surgical interventions, which may require a more aggressive approach, as the remaining neoplastic cells are usually poorly defined.
3. Wide resection is a form of curative-intent surgery: the intent is to resect macroscopic (primary tumor) and microscopic (satellite cells) tumor burdens, including biopsy tracks. This surgical dose is recommended over intralesional and marginal resections for treatment and management of solid tumors (see Determining Surgical Margins).
4. Radical resection is removal of an entire tissue compartment. Examples include removal of a complete organ, such as splenectomy to remove a splenic hemangiosarcoma or limb amputation to treat appendicular osteosarcoma. Depending on tumor location, type, and size, radical resection is occasionally needed to completely excise the primary tumor and its surrounding pseudocapsule and satellite tumor cells.

In some cases, adequate surgical margins may be too disfiguring, painful, or expensive; minimally advantageous; or simply incompatible with life. Therefore, when planning radical resections, strongly consider patient quality of life.

Determining Surgical Margins

In order to achieve a wide resection, excise a normal tissue margin en bloc with the gross tumor. Determine the width of the surgical margins by:

- Tumor type and grade, if appropriate (eg, mast cell tumors)
- Biologic behavior
- Anatomic location
- Barrier provided by surrounding tissues (tumor resistant fascial layers).

It is imperative to remember that margins are 3-dimensional:

- Lateral margin width is determined by tumor type and biologic behavior.
    - Benign tumors and most malignant carcinomas can be completely resected with 1-cm lateral margins.
    - Soft tissue sarcomas can be resected with 3-cm lateral margins.
    - Mast cell tumors are typically resected with at least 2-cm lateral margins, although grade 1 tumors may be adequately excised with 1-cm margins.

In order to accurately achieve the desired lateral margin, use of a surgical marking pen and ruler is advisable (Figure 2).

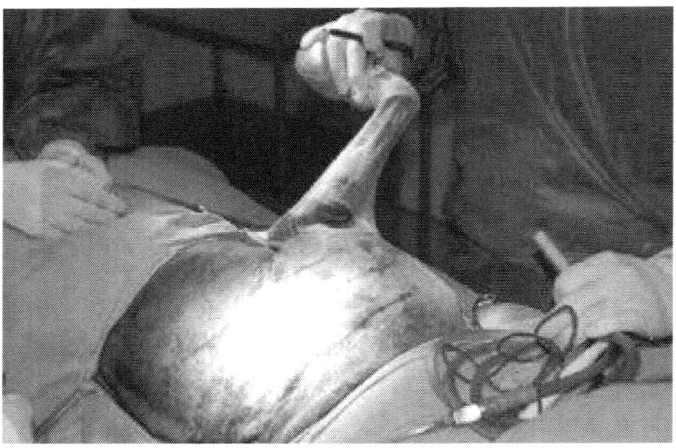

Fig. 6.2 Mark the lateral surgical margins with a surgical marking pen; for this soft-tissue sarcoma, the edge of the visible tumor is first marked, followed by marking a 3-cm lateral margin.

- Deep margins tend to be the hardest to predict, and are universally determined by natural tissue barriers if deep margins (1–3 cm) are not possible.
    - Fat, subcutaneous tissue, muscle, and parenchymal tissue do NOT provide adequate barriers to tumor invasion or suffice as deep margins (Fig. 6.3).
    - Connective tissues, including muscle fascia, cartilage, and bone, are resistant to neoplastic invasion, proving a natural tissue barrier.

Because of this protective barrier, deep margins should include removal of at least 1 fascial plane; however, vaccine-associated sarcomas require removal of at least 2 fascial planes.

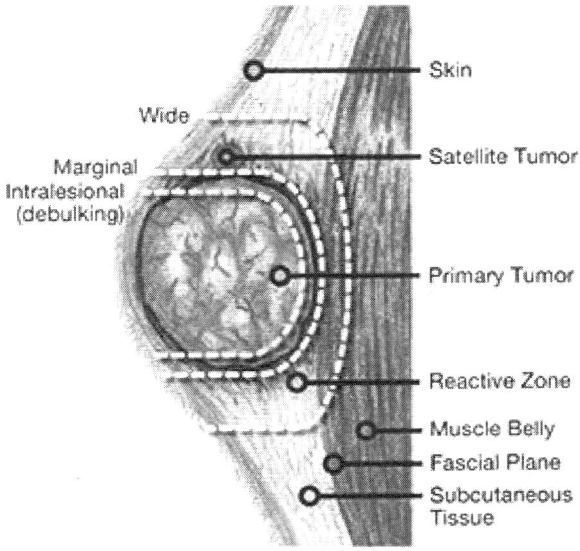

Fig. 6.3 Deep surgical margins are dependent upon the tissue layers and fascial planes, not actual measured depth. The intralesional margin is an incomplete resection inside the surrounding pseudocapsule (black line), marginal resection is dissection exterior to the surrounding pseudocapsule within the reactive zone (yellow), and wide margin resection removes both microscopic and macroscopic disease, including at least 1 fascial layer. Illustration by Michael Orencole.

TISSUE HANDLING

Careful tumor handling is required to prevent exfoliation of cells and local recurrence. To minimize seeding of neoplastic cells, avoid direct handling of the tumor and change surgical gloves and instrument packs between tumor excision and closure.

Surgical Closure

For some large defects in areas that can be easily bandaged (eg, extremity, trunk), consider leaving the wound open while awaiting the biopsy report. If the margins are not clean, resect additional tissue until a clean margin is obtained; then close the wound. This is particularly useful when the wound requires major reconstructive surgery, such as an axial pattern flap or skin graft.

Make the decision to use a surgical drain during closure carefully, and avoid if possible. Placement of a drain may increase the chance of seeding neoplastic cells and complicate future resections, if the primary resection was incomplete.

Tissue Analysis

Resected tissue should be properly prepared for analysis:

- If the resection is marginal, it is helpful to "pin" the tissue on cardboard, which prevents shrinkage during fixation.
- Marking the surgical margins with ink (eg, India ink) or suture strands helps the pathologist determine the extent of tumor resection.
- With large tumors, making several partial thickness slices (ie, "loafing") with a scalpel blade ensures the entire sample is adequately fixed.
- Tissue should be promptly placed in formalin at a ratio of 10 parts formalin to 1 part tissue.8

Dirty Margins

Despite the surgeon's best efforts, biopsy analysis following resection occasionally reveals that the tumor was inadequately resected. Treatment of choice for most dirty margins is additional surgery to excise the surgical scar and an additional margin of normal tissue (at least 1 cm). If additional resection is not feasible due to tumor location, adjuvant therapy, such as radiation, may further eliminate the local tumor burden and decrease the chance for recurrence.

General Rules for Collecting Diagnostic Tissue Biopsies

1. Substantial tissue biopsies yield better diagnostic results.
2. Multiple tissue samples increase the overall understanding of the sampled site.
3. When handling tissue samples, take great care to avoid iatrogenic artifacts.
4. Always submit tissue samples to a pathologist along with pertinent, detailed information about the patient and tissues biopsied.
5. Remove any biopsy tracks or scars during tumor excision.
6. Fix tissue samples from needle-core, incisional, and excisional biopsies in 10% buffered formalin at a ratio of 10:1 (formalin:tissue).

PRINCIPLES OF SURGICAL ONCOLOGY IN DOGS AND CATS

Oncology is a science which investigates the pathological phenomenon of uncontrolled cell proliferation in all its multifactorial aspects.

Surgical oncology, as its name suggests, is the specific application of surgical principles to the oncologic setting. These principles have been derived by adapting standard surgical approaches to the unique situations that arise when treating cancer patients.

Cancer is one of the most common causes for mortality of small animal companion animals. The role of the surgeon is important in the treatment of cancer because it often provides an immediate cure. Cure is defined in terms of the tumor-free interval the specific cancer associated with a negligible chance of recurrence. Disease-Free Interval of three years or more after treatment of most solid cancers in small animals are associated with so few clinical recurrences that is safe to assume the patient will never be bothered by that tumor again.

Surgery can be the sole source of treatment or can be combined as part of a multidisciplinary protocol including chemo, immuno, and radiation therapy. Surgery can be used to obtain a diagnosis, a complete cure of the disease, to stage the tumour, for palliation, and for debulking (cytoreduction) in combination with another treatment modality.

Diagnosis of the type of tumour can be obtained by fine needle aspiration and cytological examination. However, histological evaluation of a biopsy specimen allows a more definitive diagnosis. A biopsy specimen can be obtained using a needle or skin punch biopsy instrument or an incisional and excisional, surgical biopsy technique, biopsy techniques under endoscopic, laparoscopic, and diagnostic imaging.

General rules for biopsy

1. The tissue removed should be representative of the whole lesion.
2. Avoid zones of hemorrhage, necrosis, and obvious infection.
3. Avoid crushing and charring of removed tissues.
4. Include healthy marginal tissue: lesions of skin and mucosal surfaces.
5. Avoid massaging cancer cells into the circulation.
6. Avoid direct implantation of released malignant cells into adjacent healthy tissue.
7. Avoid repeated biopsies of the same lesion.

Fine Needle Aspiration Biopsy (Fnab)

FNAB is best performed using a 22 G. needle and a 10 ml syringe. Tumour cells are aspirated by applying suction and gentle manipulation of the needle in the tumour tissue. Cytology allows a quick diagnosis of exfoliative tumours. Mesenchymal tumours may be hard to diagnose using this technique.

Core Needle Biopsy

A core needle obtains a cylinder of tumour tissue. The needle is inserted into the tumour after the animal has been sedated. The stilet of the needle is pushed into the tumour and a core of tumour is trapped in the recess of the needle after the outer cutting layer of the needle is advanced over the stilet. The needle is then withdrawn from the tissue. Histological examination of the biopsy specimen will allow diagnosis in most tumours. Tough connective tissue tumours and blood-filled tumours may pose a problem.

INCISIONAL AND EXCISIONAL BIOPSY TECHNIQUES

Larger amounts of tissue can be obtained by incisional and excisional biopsy techniques. These techniques will allow better evaluation of tumour characteristics, including margins, but require general anesthesia. Excisional biopsies are performed in areas where the tumour can be easily removed using adequate margins or in case of benign lesions. Incisional biopsy techniques are best used for large tumours, or for tumours located in places where wide excision is not possible. The location of the biopsy incision is of extreme importance, because it needs to be excised during later surgical procedures.

The usual indication for biopsy of the lymph node is to establish the diagnosis of lymphoma or metastatic carcinoma. Each situation should be approached in a different manner.

The goal of biopsy in the patient with an abnormal lymph node and suspected lymphoma is to make the general diagnosis and to establish the lymphoma type. Consequently, the initial diagnosis of lymphoma should be made on a completely excised node that has been minimally manipulated to ensure that there is little crush damage. When primary lymphoma is suspected, the use of needle aspiration does not consistently allow for the complete analyses described above and can lead to incomplete or inaccurate diagnosis and treatment delays.

The diagnosis of metastatic carcinoma often requires less tissue than is needed for lymphoma. Fine-needle aspiration (FNA), core biopsy, or subtotal removal of a single node will be adequate in this situation. For metastatic disease, the surgeon will use a combination of factors, such as location of the node, physical examination, and symptoms, to predict the site of primary disease. When this information is communicated to the pathologist, the pathologic evaluation can be focused on the most likely sites so as to obtain the highest diagnostic yield. The use of immunocytochemical analyses can be successful in defining the primary site, even on small amounts of tissue.

For soft-tissue or bony masses of the trunk or extremities, the biopsy technique should be selected on the basis of the planned subsequent tumor resection. The incision should be made along anatomic lines in the trunk or along the long axis of the extremity. When a sarcoma is suspected, FNA can establish the diagnosis of malignancy, but a core biopsy will likely be required to determine the histological type and plan neoadjuvant therapy.

The surgical planning for the operative procedure should be made carefully. The best, and often the only, opportunity for cure is at the time of the first surgery. Enucleation or incomplete excision of tumor masses is never indicated as a therapeutic measure.

There are major determinants of operative risks: General health status, severity of underlying illness, degree to which surgery disrupts normal physiologic functions, technical complexity of the procedure (related to incidence of complications), types of anesthesia required, and experience of the personnel.

Preoperative preparation of the patient should consider that tumor acts as a nitrogen trap because of its rapid growth as the metabolic expense of the rest of the body. It is important to evaluate serum albumin and if necessary give IV hyperalimentation. Particularly in gastrointestinal cancers due to bleeding anemia is often present and as necessary measure blood volume and replace blood or red cells deficit. In the elderly patient associated cardiovascular and pulmonary disease may be present, which should be treated preoperatively guided by thoracic radiographs, ECG, BUN, and electrolytes if vomiting is present. Rule out obvious possible sites of metastases such as liver, lung and bones where indicated by checking skeletal and thoracic radiographs, abdominal US and liver chemistries.

The extent of surgery depends on the type and stage of the tumour. A pre-resection biopsy is often necessary to anticipate sufficient margins of resection and behavior of the primary tumour. Staging the tumour will prevent treatment failures caused by early metastasis and will help in deciding the type of resection and use of adjuvant treatment modalities. For some tumour types, histological grading may be valuable to anticipate the prognosis after surgery. Diagnosis of intercurrent diseases often will alter surgical treatment options and should be evaluated beforehand to assess the risk versus benefit of surgical intervention. The surgical field should be prepared carefully to allow changes in the extent of resection based on new information obtained during the operation. The risk of microfocal, regional or distant dissemination generally increases with the advancing clinical and pathological stage of disease. Later the disseminated phase is reached, better the diagnosis. This concept indicates the need to attack a cancer during its local and regional phases providing "wide local excision" with resection of the primary lymphatic drainage

Surgical Technique

Oncologic surgery often combines many surgical specialties. Knowledge of general, orthopedic, reconstructive, and oncologic surgical techniques will need to be combined with a solid anatomical knowledge base to prevent unpleasant and unnecessary surprises. Dissection of tumours should be clean and exposure of the tumour should be prevented. Tumour cell contamination (seeding) is minimized by avoiding tumour incision, shielding normal tissue by protective covering, and using proper electrosurgical techniques. In case of contamination, contaminated sites should be resected (where possible), the wound should be lavaged, and contaminated gloves, drapes and instruments replaced. Adjunctive therapy often is indicated in these cases. Tissue handling should be according to Halstead's principles and the tumour itself should be handled as little as possible. Neovascularization is often prominent and should be dealt with properly by electrocoagulation or suture ligation. Tissues macroscopically invaded by the primary tumour should be excised 'en bloc'.

SURGERY FOR CURE

Resection for cure must be aggressive and at an early stage of disease. The most important theme for oncologic surgical management is that the first time has the best chance of curing the patient. Margins should not be compromised because it is better to leave a wound open to heal by second intention than to leave tumour cells behind. Recurrent tumours are always more difficult to treat than the primary ones. Familiarity with tissue grafts, skin flaps and other reconstructive techniques will allow wider resections, minimize patient morbidity (and in the end patient mortality) and decrease the temptation to compromise the resection, and thus recurrence.

SURGERY FOR PALLIATION

The goal of palliative surgery is to improve the quality of the patient's life, without the guarantee that it will increase the survival time. This type of surgery demands a very careful consideration of the risk versus benefits but will be beneficial for certain patients (e.g., amputation of an extremities face a metastatic osteosarcoma).

SURGERY FOR CYTOREDUCTION

Incomplete removal of the tumour is only beneficial in cases in which total resection is not possible, in which adjunctive therapy may benefit from the procedure, and in cases of severe morbidity caused by the primary tumour. Cytoreduction is, however, rarely indicated and should be avoided.

Surgical Margins

Surgical margins are dictated by the tumour type and grade, and the anatomical location of the tumour. The most rational approach is to think of biologic rather than geometric margins and to combine this information with the expected growth behavior of the given tumour type. Tumours with high probability of local recurrence (e.g., mast cell tumours, feline mammary tumours) should have 2 to 3 cm margins removed three-dimensionally. Collagen/matrix-rich and poorly vascularized tissues are least vulnerable to tumour invasion and may be used for margin determination. All previously performed biopsy tracts should be removed in continuity with the primary tumour to prevent tumour seeding. With a proper resection the tumour is never visualized.

Postoperative Care

Supportive care is provided as necessary for vital organ function, nutritional support, and wound healing. Appropriate pain management is essential and often neglected. Resected tumour specimens should be sent for histological examination to evaluate tumour type, grade, and surgical margins. Questionable results are discussed with the pathologist. Long-term monitoring includes regular examinations to facilitate early detection of tumor recurrence. Follow up do includes rehabilitation by different physiotherapy techniques.

CONCLUSION

In veterinary medicine 80% of the cases of companion animal cancers are treated by surgery and surgery is employed in 100% of diagnostic and statement situations. Correct and aggressive surgical resection after a solid diagnosis in an early stage of the disease will be associated with the best results.

7

ABDOMINAL SURGERY

The term abdominal surgery broadly covers surgical procedures that involve opening the abdomen (laparotomy). Surgery of each abdominal organ is dealt with separately in connection with the description of that organ (see stomach, kidney, liver, etc.) Diseases affecting the abdominal cavity are dealt with generally under their own names (e.g. appendicitis).

TYPES

The most common abdominal surgeries are described below.

- **Appendectomy**—Surgical opening of the abdominal cavity and removal of the appendix. Typically performed as definitive treatment for appendicitis, although sometimes the appendix is prophylactically removed incidental to another abdominal procedure.
- **Caesarean section** (also known as C-section) is a surgical procedure in which one or more incisions are made through a mother's abdomen (laparotomy) and uterus (hysterotomy) to deliver one or more babies, or, rarely, to remove a dead fetus.
- **Inguinal hernia surgery**—This refers to the repair of an inguinal hernia.
- **Exploratory laparotomy**—This refers to the opening of the abdominal cavity for direct examination of its contents, for example, to locate a source of bleeding or trauma. It may or may not be followed by repair or removal of the primary problem.
- **Laparoscopy**—A minimally invasive approach to abdominal surgery where rigid tubes are inserted through small incisions into the abdominal cavity. The tubes allow introduction of a small camera, surgical instruments,

and gases into the cavity for direct or indirect visualization and treatment of the abdomen. The abdomen is inflated with carbon dioxide gas to facilitate visualization and, often, a small video camera is used to show the procedure on a monitor in the operating room. The surgeon manipulates instruments within the abdominal cavity to perform procedures such as cholecystectomy (gallbladder removal), the most common laparoscopic procedure. The laparoscopic method speeds recovery time and reduces blood loss and infection as compared to the traditional "open" method.

Complications

Complications of abdominal surgery include, but are not limited to:

- Adhesions (also called scar tissue): Complications of postoperative adhesion formation are frequent, they have a large negative effect on patients' health, and increase workload in clinical practice
- Bleeding
- Infection
- Paralytic ileus (sometimes called ileus): short-term paralysis of the bowel
- Perioperative mortality (death)
- Shock

Sterile technique, aseptic post-operative care, antibiotics, use of the WHO Surgical Safety Checklist, and vigilant post-operative monitoring greatly reduce the risk of these complications. Planned surgery performed under sterile conditions is much less risky than that performed under emergency or unsterile conditions. The contents of the bowel are unsterile, and thus leakage of bowel contents, as from trauma, substantially increases the risk of infection.

Globally, there are few studies comparing perioperative mortality following abdominal surgery across different health systems. One major prospective study of 10,745 adult patients undergoing emergency laparotomy from 357 centres in 58 high-, middle-, and low-income countries found that mortality is three times higher in low- compared with high-HDI countries even when adjusted for prognostic factors. In this study the overall global mortality rate was 1•6 per cent at 24 hours (high 1•1 per cent, middle 1•9 per cent, low 3•4 per cent; $P < 0•001$), increasing to 5•4 per cent by 30 days (high 4•5 per cent, middle 6•0 per cent, low 8•6 per cent; $P < 0•001$). Of the 578 patients who died, 404 (69•9 per cent) did so between 24 h and 30 days following surgery (high 74•2 per cent, middle 68•8 per cent, low 60•5 per cent). Patient safety factors were suggested to play an important role, with use of the WHO Surgical Safety Checklist associated with reduced mortality at 30 days.

Taking a similar approach, a unique global study of 1,409 children undergoing emergency laparotomy from 253 centres in 43 countries showed that adjusted mortality in children following surgery may be as high as 7 times greater in low-HDI and middle-HDI countries compared with high-HDI countries, translating to 40 excess deaths per 1000 procedures performed in these settings. Internationally, the most common operations performed were appendectomy, small bowel resection, pyloromyotomy and correction of intussusception. After adjustment for patient and hospital risk factors, child mortality at 30 days was significantly higher in low-HDI (adjusted OR 7.14 (95% CI 2.52 to 20.23), $p<0.001$) and middle-HDI (4.42 (1.44 to 13.56), $p=0.009$) countries compared with high-HDI countries.

Absorption of drugs administered orally was shown to be significantly affected following abdominal surgery.

There is low-certainty evidence that there is no difference between using scalpel and electrosurgery in infection rates during major abdominal surgeries.

CELIOTOMY is commonly performed in veterinary practice; therefore, a systematic approach to the abdominal cavity is essential for every surgical-minded vet.

The term laparotomy refers to a flank incision while celiotomy is a more correct definition of a generic surgical approach to the abdomen: in reality both are used synonymously.

Indications

Celiotomy may be performed for a variety of reasons, both therapeutic and diagnostic.

- Acute onset abdominal pain and findings suggestive of life-threatening abdominal pathology (GDV, intestinal obstructions and volvulus, etc.).
- Abdominal trauma/foreign bodies.
- Septic/aseptic peritonitis.
- Haemoabdomen/internal haemorrhage in haemodynamic unstable patient.
- Uroabdomen.
- Pneumoabdomen.
- Chyloabdomen.

- Biopsies of abdominal organs (chronic pathologies of abdominal organs, staging of neoplasia, etc.).
- Approach to the retroperitoneal space.
- Reconstruction of abdominal wall defects and hernias.
- Specialised surgery of the gastrointestinal tract, liver and pancreas, urinary system, reproductive system, haemolymphatic system.

Contraindications

All patients undergoing a celiotomy must be haemodynamically stable enough to go through anaesthesia. Therefore, compromised patients must be adequately stabilised before surgery.

While an explorative laparotomy represents a useful diagnostic tool, it must be performed only when extra-abdominal conditions have been ruled out. Inadequate surgical training and equipment also represent contraindications to surgery.

Pre-operative patient management

Pre-operative management depends on the underlying pathology. General indications are: performing an accurate general examination with special attention to the cardiovascular system; adequate diagnostic imaging work-up according to the initial presentation; placing of an intravenous catheter; blood works including a minimum of PCV, CBC, platelet count, total proteins, BUN and blood glucose; more biochemistry and clotting profiles according to the animal disease or if a coagulopathy is suspected; blood typing if an intra-operative haemorrhage is foreseen.

ANAESTHETIC/ANALGESIC CONSIDERATIONS

An anaesthetic plan must be organised according to the signalment, history, underlying disease process and ASA status classification. Appropriate analgesia is paramount. Pre-emptive and intra-operative analgesia may include opioids, ketamine, alpha-two agonists and local anaesthetics. Postoperative analgesia may include also NSAIDs.

Anaesthetic monitoring such as ECG, pulse oximetery, non-invasive or invasive blood pressure monitoring, capnography and body temperature monitoring are vital for debilitated patients. Mechanical ventilation may be required.

Surgical anatomy of the abdominal wall

The abdominal ventral wall is composed of an external and an internal leaf. The external leaf is formed by the aponeurosis of the external and internal abdominal oblique muscles and by the aponeurosis of the transversus abdominis muscle near the pubis.

The internal leaf is composed by a portion of the aponeurosis of internal abdominal oblique muscle, by the aponeurosis of the transversus abdominis muscle and by the transversal fascia. The internal leaf disappears in the caudal third of the abdomen.

Surgical instrumentation

A basic laparotomy tray should include: scalpel handle for 10 and 11 blades, needle holder (Mayo or Mathieu), Brown-Adson, DeBakey and thumb forceps, Allis tissue forceps, Babcock forceps, Halstead Mosquito and Rochester-Carmalt forceps (straight and curved), Metzenbaum and Mayo scissors (straight and curved), Doyen intestinal clamps, Balfour or Gosset self-retaining retractors, Senn-Miller or Faraboeuf manual retractor.

Also necessary are large laparotomy sponges, large and small swabs with radio-opaque markers, electrocoagulation, suction and appropriate suture material. Sterile water-per injection or Hartmann's must also be available to be warmed to flush the abdominal cavity if needed. Surgical instrumentation must be organised on the surgical tray before the first incision.

PATIENT PREPARATION

The patient is placed in dorsal recumbence. The entire abdomen, including the inguinal area, the medial side of the legs and the caudal thorax must be clipped. The area clipped must be large enough to accommodate unforeseen complications and for expansion of the field if necessary.

The prepuce can be retracted to one side with a towel clamp but if left in the surgical site it must be flushed with an appropriate diluted antiseptic solution. The bladder should be voided prior to surgery and a urine sample obtained.

An initial layer of four drapes must be placed around the scrubbed area, and then a second bigger drape will cover all the exposed portion of the patient and the surgical table.

It is also possible to apply an antimicrobial adhesive drape to the site of the incision: this helps prevent leaking of fluid around the drapes.

All the swabs and sponges must be counted; they will be recounted before closure of the abdomen to verify that none has been left in the abdominal cavity.

Surgical approach to the abdominal cavity

- Incise the skin on the ventral midline with a 10 blade scalpel, going from xiphoid process and extending caudally to the pubis. The incision must be long enough to permit a complete exploration of the abdominal cavity. In male dogs extend the incision around the prepuce cutting only through the skin to avoid damage to the external pudendal vessels. It is not possible to perform an exploratory laparotomy through a small excision.
- With Metzenbaum scissors extend the incision to the subcutaneous fat. In cats the linea alba is easily identified once the subcutaneous fat has been trimmed. In dogs subcutaneous fat can obscure linea alba, therefore more dissection may be necessary (push-cut technique). Ligate or cauterise small subcutaneous vessels.
- In the male dog, identify and transect the preputial muscle (the two ends of this muscle will be re-apposed during closure). Identify, ligate and transect the superficial branches of the pudendal muscle.
- With the linea alba visible, tent it away from abdominal organs with thumb forceps. Holding the scalpel with the cutting edge facing up and parallel to the body of the patient, make a sharp incision into the linea.
- With fingers or closed blunt forceps, palpate the interior surface cranially and caudally to verify there are no adhesions.
- Use Mayo curved scissors or a blade to extend cranially and caudally the linea alba incision.
- The falciform ligament cranially must be removed if it interferes with a complete exploration of the abdominal cavity. Its lateral attachments are transected and the base is ligated and removed.
- Gosset retractors are essential to maintain exposure of the abdomen. Place moistened laparotomy sponges between the arms of the retractors and abdominal wall to protect tissue from excessive pressure, making sure no organs are caught in them.

Systematic exploration of the abdominal cavity

Systematic evaluation of all the abdominal contents must be performed in all cases. If fluid is present, samples must be obtained for evaluation.

If present, isolation and control of internal haemorrhage and gastrointestinal leakage must take priority. All organs must be gently manipulated and maintained moistened with laparotomy sponges or warm solution.

The abdominal cavity is divided into four quadrants and thoroughly explored. All organs are examined, paying attention to size, shape, colour, location and consistency. Normally the first thing seen on opening the abdomen is the greater omentum attached to the greater curvature of the stomach and covering the abdominal organs.

- **Cranial quadrant:** exteriorise, inspect and palpate the two leafs of the greater omentum. With gentle traction, displace the liver and examine the diaphragm and the hiata. Evaluate and palpate all liver lobes. Inspect and express the gall bladder and check the integrity of the extra-hepatic biliary tree. Examine and palpate the whole stomach, the mobilisation of which is limited by the oesophagus and the pylorus. Inspect and palpate the duodenum and the duodeno-pancreatic vessels. The spleen can be examined in situ or completely exteriorised; check both surfaces for pulsation of the splenic artery. Inspect both the pancreatic limbs, handling it gently; to visualise the left, apply a gentle traction on the body of the stomach. Then examine the portal vein, the caudal vena cava and the hepatic arteries.
- **Caudal quadrant:** inspect the descending colon and mesocolon. The urinary bladder should be emptied prior to surgery; applying traction to its pole, it's possible to visualise also its dorsal face. Digitally, palpate and examine the intrapelvic urethra. By manipulating the urethra it's possible to examine both lobes of the prostate. The uterine body and horns are deep in the abdominal cavity, in an extraomental position. Inspect the inguinal rings and the local vascularisation. Examine the rectum by applying gentle caudal traction on the descending colon.
- **Intestinal tract:** the greater omentum is displaced cranially and the intestinal tract is examined starting from the duodenum. The jejunum and the ileum are elevated outside the abdominal cavity and palpated. Check the mesenteric vascularisation and lymph nodes. Mesenteric arterial pulsation and intestinal peristalsis must also be evaluated.
- **"Gutters":** use the mesoduodenum as an anatomical retractor to displace the intestines medially (left side of the animal) and examine the right abdominal contents. Palpate the kidneys and examine the adrenal glands, the ureter and the ovary if present. Inspect the renal artery and veins, following them till the caudal vena cava. Similarly, using the descending colon to retract the abdominal contents to the right side of the animal, inspect and palpate the left kidney, adrenal gland, ureter and ovary.

- Part two will include indications and instructions as to how to obtain biopsies in the abdomen and closure of the abdomen.

Preoperative Preparation

Celiotomy is performed for various reasons; it may be indicated for diagnostic (e.g., biopsying an organ) and/or therapeutic reasons. Many animals undergoing abdominal exploratory surgery have chronic disease, but some patients require emergency abdominal surgery because of acute clinical signs. Some conditions are life-threatening (e.g., gastric dilatation-volvulus, colonic perforation, severe hemorrhage), and appropriate therapy must be started promptly. Conditions that require surgery must be differentiated from those that can be managed medically. Although obviously unnecessary surgery must be avoided, surgery cannot always be delayed until one is certain the patient will benefit from it.

The decision to operate is based on the history and physical examination findings, radiographic and ultrasonographic studies, and laboratory analyses. Physical examination can be unreliable in predicting the severity of abdominal trauma. The inaccuracy associated with examining patients with acute abdominal disease, particularly that associated with trauma, can be attributed in part to the patient's condition at the time of examination and delayed development of clinical signs associated with some injuries. Depressed or lethargic animals may not show pain during abdominal palpation. Clinical signs of hemorrhage often are inapparent immediately after trauma; delays of 3 to 4 hours between injury and development of shock and collapse are common in patients with hepatic or splenic lacerations. Therefore, animals that have suffered traumatic injury should be closely observed for at least 8 to 12 hours. Life-threatening hemorrhage becomes apparent before this time in most cases. However, animals with traumatic bilious peritonitis may not have overt clinical signs for weeks. Likewise, traumatic mesenteric avulsion is seldom associated with clinical signs until peritonitis develops, usually several days after injury. Sensitive diagnostic tests such as diagnostic peritoneal lavage may help identify patients with significant abdominal trauma before overt clinical signs develop.

Preoperative management of most animals undergoing exploratory laparotomy is dictated by the underlying abdominal disease. General observations include noting the animal's attitude and posture, temperature, respiratory rate and effort, and heart rate and rhythm. Abdominal auscultation, percussion, and palpation plus rectal examination are indicated. Serial examinations are important to detect trends or deterioration in the patient's status. An intravenous catheter should be placed for fluid and drug administration, and blood samples should be drawn.

Useful initial blood work in an animal with acute abdomen includes complete blood count (CBC), platelet count, serum total protein and glucose concentrations, and blood urea nitrogen (BUN). Other laboratory tests (e.g., serum biochemistry profile, clotting parameters) can be performed, depending on the animal's condition and the suspected underlying disease. Urine may be collected by means of cystocentesis or catheterization for urinalysis. An indwelling urinary catheter may be used to quantitate urinary output if necessary. Abdominal radiographs may detect peritoneal fluid (i.e., uroabdomen, peritonitis) or abnormal accumulations of air. Animals with acute abdominal signs of uncertain cause should have abdominocentesis, diagnostic peritoneal lavage (see p. 380) FAST (see p. 383), or a CT performed if radiographs are nondiagnostic. Electrolyte and hydration abnormalities should be corrected before surgery.

Major abdominal evisceration may occur in dogs secondary to postsurgical dehiscence or trauma. In affected animals, the intestines typically are eviscerated and may have gross contamination with dirt or other debris (e.g., cat litter). Regardless of the inciting cause, exposure and contamination of the abdominal viscera warrant immediate surgical intervention. In a recent study of animals that had developed a postoperative major abdominal evisceration, all had recently undergone ovariohysterectomy (Gower et al, 2009). The authors noted that the preponderance of wound dehiscence leading to abdominal evisceration in animals undergoing ovariohysterectomy was likely a reflection that it is one of the most frequently performed abdominal procedures in small animals, rather than being a direct consequence of the procedure.

Perioperative antimicrobials are administered to animals with existing infection or prophylactically for surgical wounds classified as clean-contaminated, contaminated, or dirty. Prophylactic antimicrobials are given intravenously 30 to 60 minutes before incision and may be repeated during the procedure, depending on surgical duration, predicted contaminants, and the drug's metabolism in that species. In dogs, an intramuscular or subcutaneous dose of cefazolin can be given concurrently with the intravenous dose to provide extended prophylaxis.

The patient's entire abdomen and caudal thorax are clipped and aseptically prepared. In male dogs, the preputial cavity is flushed with saline or water before application of antiseptic solution, since antiseptics are inactivated by organic material. For female dogs with suspected urinary tract obstruction, the perivulvar region is also clipped and prepped so that it can be draped in for intraoperative catheterization. Before beginning the procedure, the surgeon and anesthetist or veterinary nurse should review a surgical checklist that includes preoperative and intraoperative medications, a surgical sponge count, and a list of surgical and anesthetic concerns.

Abdominal Incision

For a full exploratory laparotomy, the abdomen is incised from xiphoid to pubis. In male dogs, the skin and subcutaneous incisions deviate laterally around the prepuce; branches of the caudal superficial epigastric vessels may require ligation or cauterization. Once the subcutaneous tissues are incised, the abdominal cavity is entered through an incision in the linea alba (Fig. 7.1). If the linea is difficult to identify, the subcutaneous tissues can be sharply transected with scissors at their midline attachments (Fig. 7.2).

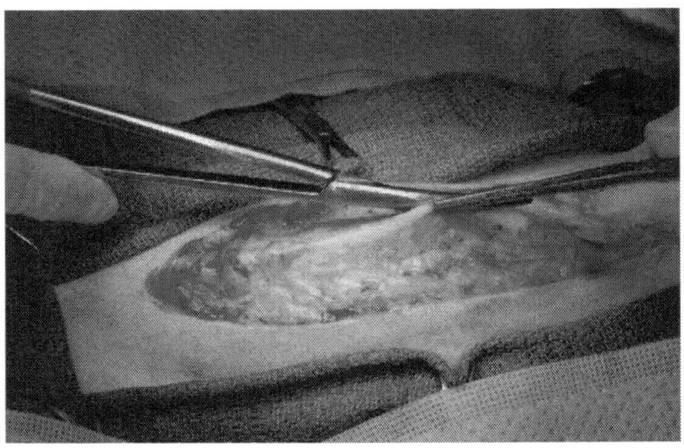

Fig. 7.1 Initial incision into the abdomen through the linea alba.

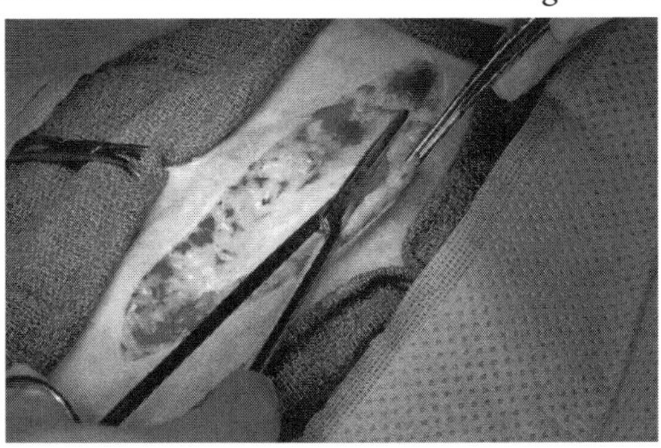

Fig. 7.2 Clearing the subcutaneous tissues away from the linea alba.

The falciform ligament often obscures visualization of cranial abdominal components and requires lateral transection and cranial ligation. A self-retaining retractor (e.g., Balfour) helps maintain visualization, particularly of the cranial abdomen; the surgeon should verify there are no organs trapped between its lateral blades and the abdominal wall. Moistened laparotomy sponges can be placed underneath retractor blades to protect underlying skin and muscle.

Once the abdomen is open, the presence, quantity, and quality of fluid should be assessed. If septic effusion is encountered, samples for culture and Gram stain should be obtained. For patients with hemoperitoneum, blood loss can be calculated by adding the amount of blood in the suction canister to the estimated amount of gauze saturated with blood. When completely soaked, a 4 × 4 gauze pad will hold approximately 12 mL of blood, while a 30 cm × 30 cm laparotomy sponge will hold approximately 100 mL of blood.

EVALUATION

An appropriate exploratory laparotomy relies on the veterinarian's knowledge of anatomy (Fig. 7.3) and attention to detail. To avoid missing lesions, a consistent, systematic technique is used. One common method is to sequentially examine structures in the cranial abdomen (diaphragm, liver, gallbladder, stomach, pylorus, proximal duodenum, spleen), right and left gutters (descending duodenum, pancreas, kidneys, adrenal glands, ovaries, descending colon), and caudal abdomen (urinary bladder, ureters, uterus, urethra, prostate, sublumbar lymph nodes), before "running" the intestines (i.e., examining and palpating them for lesions). If lesions are not found, the surgeon should be prepared to take a variety of samples for diagnosis, particularly in patients with digestive tract signs.

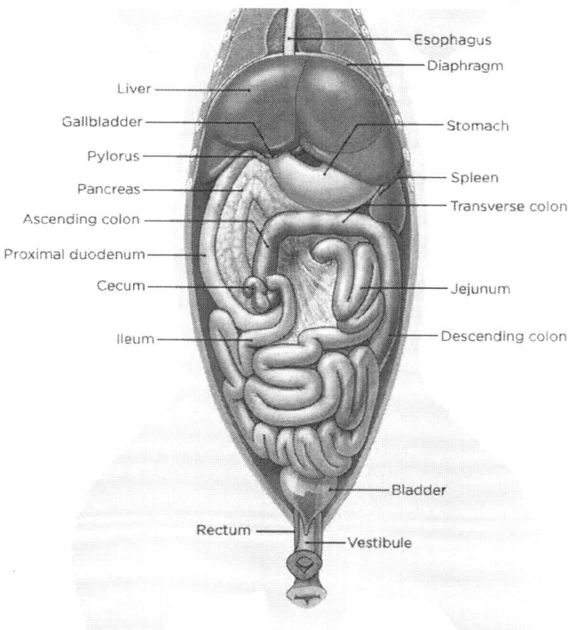

Fig. 7.3(a) Ventral view of abdominal contents in a female dog. (A) Normal location of organs. Illustration: Kip Carter.

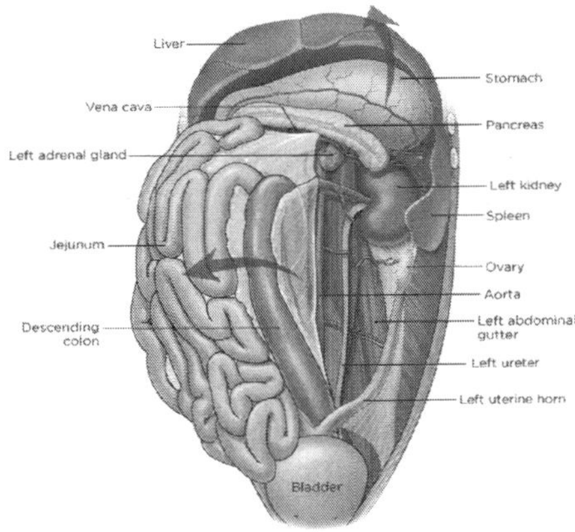

Fig. 7.3(b) Intestines retracted to show dorsally located organs. Illustration: Kip Carter.

Diaphragm

The diaphragmatic central tendon and muscular pars lateralis and costalis are normally taut and concave. Flaccidity or loss of concavity can indicate the presence of pneumothorax, pleural effusion, intrathoracic masses, or diaphragmatic hernia. Any of these findings should trigger a request for immediate ventilation monitoring and support. Diaphragmatic visualization requires hepatic, gastrointestinal, and splenic retraction.

LIVER AND GALLBLADDER

The liver consists of left lateral, left medial, quadrate, right medial, right lateral, and caudate lobes. The gallbladder lies between the quadrate and right medial liver lobes; the apical half is visible, while the remainder is adhered to surrounding hepatic tissue. Normal liver lobes are dark red, smooth, and sharply marginated. A normal gallbladder is oval or round, thin walled, and easily expressible in dogs. Because the common bile duct and pancreatic duct in cats may be conjoined or open in close proximity, gallbladder expression is not recommended in cats. Hepatic abnormalities include changes in size, texture, or color; marginal rounding; and bleeding or mass-like lesions. Abnormal gallbladders are often firm, distended, discolored, or adhered to other organs.

For generalized or peripheral liver disease, a guillotine biopsy can be performed by ligating a portion of the lobe margin with a loop of absorbable suture tied with a single surgeon's throw (Fig. 7.4). Hepatic tissue is excised distal to the ligature, which is left in place. Likelihood of obtaining diagnostic samples is increased when multiple liver lobes are biopsied.8 Focal or central lesions are sampled with a skin biopsy punch. If hemorrhage persists after sample removal, a clotting adjuvant can be applied to the defect.

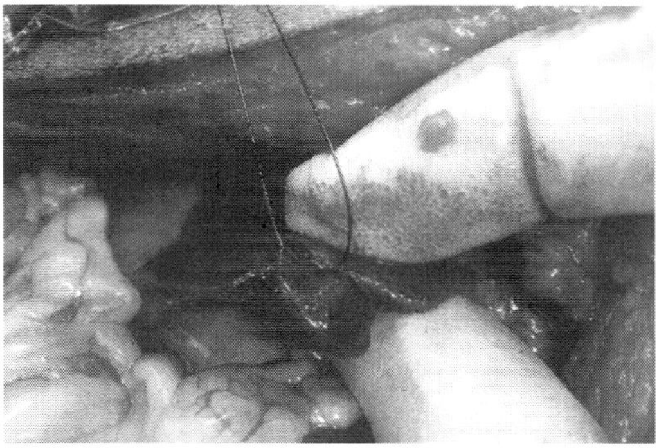

Fig. 7.4 Guillotine biopsy of the liver. Courtesy Josep Aisa, DVM, DECVS, University of Tennessee.

If hepatobiliary infection is suspected, cholecentesis can be performed in conjunction with liver biopsies and cultures. If bile is thin, a 25-gauge needle may be sufficient for gallbladder aspiration. The needle can be inserted through adherent hepatic tissue to reduce leakage; if inserted through the gallbladder apex, the gallbladder should be fully drained to decrease the risk of bile leakage. Cholecentesis should be avoided in patients with biliary mucoceles, questionable gallbladder wall integrity, or complete biliary tract obstruction.

Stomach

The stomach consists of the cardia, fundus, body, and pylorus. The dorsal aspect of the stomach is viewed by making a hole in the ventral leaf of the greater omentum to enter the omental bursa. The left limb of the pancreas and splenic vasculature can be examined through the omental opening at the same time. The only firm portion of the stomach is the pylorus. The stomach should be assessed for abnormal coloration, thickness, or vascular pattern. Biopsy is performed via a full-thickness gastrotomy. For surgeons unfamiliar with normal pylorus size and thickness, pyloric assessment may require gastrotomy and digital insertion.

Spleen

The spleen lies in close association with the greater curvature of the stomach and is attached to it by the gastrosplenic ligament. A normal spleen is dark purple in color, with smooth, rounded edges, but size and color can vary depending on the degree of splenic contraction and effect of anesthetic drugs. The spleen is evaluated for changes in size, symmetry, and texture; capsular integrity; and vascular supply (e.g., tears, thrombosis, torsion). Although splenic biopsy can be performed using a guillotine technique or wedge incision and closure, the spleen is usually completely removed if neoplasia, thrombosis, or infection is suspected.

Descending Duodenum and Right Limb of the Pancreas

The descending duodenum is the most dorsal intestinal structure in the right gutter and is easily identified by the associated right limb of the pancreas. Blood supply to the descending duodenum forms an arcade from the anastomosis of the cranio- and caudo-pancreaticoduodenal branches of the cranial mesenteric artery; these vessels often traverse through the right limb of the pancreas. The body of the pancreas sits adjacent to the pylorus, while the left pancreatic limb lies within the greater omentum along the splenic artery and vein and adjacent to the stomach and transverse colon. The common bile duct enters the serosal surface of the descending duodenum and travels intramurally about a centimeter before opening intraluminally at the major duodenal papilla. In animals with common bile duct dilation, the surrounding pancreas should be evaluated for evidence of inflammation or scarring, and the free and intramural portions of the duct can be gently palpated for obstructive choledocholiths or masses.

Pancreatic lobar tissue is normally white to pale pink, soft, thin, and pliable. The tissue can be gently palpated for nodules if an insulinoma is suspected; however, extensive or rough palpation could lead to pancreatic ischemia or inflammation. Pancreatic biopsy can be performed via blunt dissection of lobules or ligation of a free edge using a guillotine technique. Avoid sampling areas that contain major vessels or pancreatic ducts; if the pancreaticoduodenal vessels are damaged, blood supply to the duodenum could be interrupted.

Kidneys and Adrenal Glands

The kidneys lie within the retroperitoneal space and are often surrounded by fat. Both kidneys should be roughly the same size and shape and have a smooth exterior. Because of their retroperitoneal position and surrounding fat, the ureters are usually not visible as they exit the renal pelvis. Cortical renal biopsies are most often performed with a 16-gauge automated biopsy needle; for best results, 2 core samples are recommended. Renal parenchyma bleeds extensively; however, hemorrhage from needle biopsy will usually subside with local digital pressure.

The adrenal glands are elongated, whitish-tan, and firm. The left gland lies medial to the cranial pole of the left kidney and is usually visible lateral to the caudal vena cava and under the left phrenicoabdominal vein. The right adrenal gland is more cranial and may lie dorsal to the caudal vena cava, making it more difficult to see. The glands can be examined and palpated for enlargement or nodules, but because of adjacent vascular structures, sampling is not recommended unless the surgeon has advanced training.

Ovaries and Uterine Horns

In intact females, the ovaries and proximal uterine horns lie caudodorsal to the kidneys. They are usually located by retracting the descending duodenum or descending colon and associated viscera to the animal's left or right, respectively, to visualize the "gutters" of the abdominal cavity. Normal ovaries and uterus vary in size and shape depending on the reproductive stage. They should be examined for masses and cysts. In ovariectomized animals displaying signs of estrus, peritoneal tissue caudal to the kidneys and adjacent intestinal mesentery or omentum should be examined and palpated for nodules; these will be more obvious if the animal is in heat. Ovarian remnants may overlie the ureters and must therefore be dissected out carefully.

Caudal Urogenital Tract

A normal, distended urinary bladder is thin-walled, smooth, and relatively vascular. If the bladder is enlarged and obstructing visibility, it can be emptied by manual expression, cystocentesis, or catherization. The decompressed bladder can be retracted caudoventrally out of the abdominal cavity to expose the ureters, which enter at the trigone, and underlying dorsal structures such as uterine body/cervix, descending colon, and sublumbar lymph nodes.

The bladder narrows down into the trigone and proximal urethra, which is obscured by the prostate in male dogs. The canine prostate is normally bilobed and can be quite small and firm in castrated dogs. It should be examined for abnormal enlargement, fluid pockets (cysts or abscesses), or masses. Lateral to the bladder, trigone, and urethra, the vas deferens and testicular vessels become closely associated as they descend toward the inguinal canals. A cryptorchid testicle can be located by following the testicular artery and vein from the level of the kidneys or the vas deferens from the prostate.

BODY WALL

While the abdominal gutters are exposed or the bladder exteriorized, the dorsal and lateral peritoneal surfaces of the body wall can be examined. Distortion or disruption may be an indication of trauma or retroperitoneal hemorrhage, abscessation, or masses (Fig. 7.5).

Fig. 7.5 Laceration of the transversus abdominus.

Gastrointestinal Tract

Although various portions of the gastrointestinal tract will already have been examined visually, a final examination is performed by palpating the entire tract for changes in surface texture, color, diameter, thickness, and peristalsis (Fig. 7.6). The jejunum is the longest, most mobile portion of the small intestine and the most likely region to be obstructed by intestinal foreign bodies. Peyer's patches are clusters of lymphoid follicles located within the small intestine; these should not be misinterpreted as neoplastic lesions.

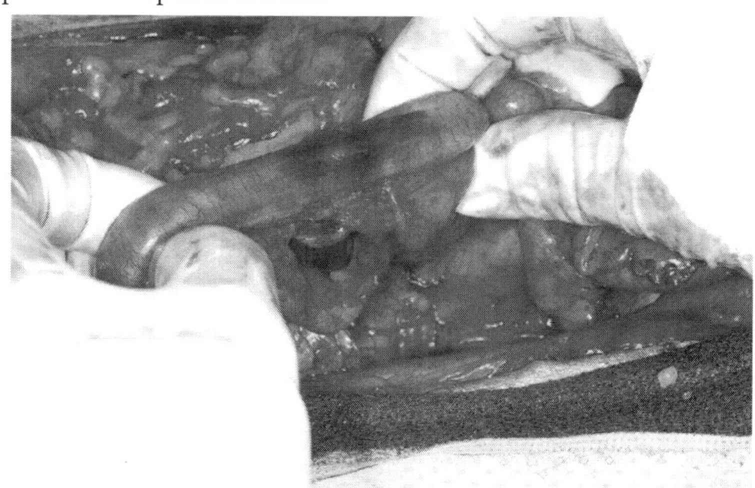

Fig. 7.6 Bite wound of the small intestine with secondary mesenteric laceration.

The ileum is easily identified by its antimesenteric branches of the ileocecal artery and vein. The adjacent cecum is a C-shaped, gas-filled protrusion. The ileocecocolic junction is another common location for foreign body entrapment. The colon consists of ascending, transverse, and descending portions, with the descending colon continuing as the rectum at about the level of the pelvic brim.

Intestinal biopsy can be performed by using a scalpel blade or skin biopsy punch to remove a full-thickness piece of tissue (Fig. 7.7). Intestinal closure requires precise apposition to prevent postoperative leakage or dehiscence; most commonly, absorbable 3-0 or 4-0 monofilament suture material is used. After closure, the segment of intestine can be checked for leakage by occluding the lumen orad and aborad to the site and distending the enclosed segment with sterile saline injected through a 25-gauge needle. When testing a 10-cm intestinal segment, 16 to 19 mL of saline should be injected if digital occlusion is used, whereas 12 to 15 mL of saline is injected if Doyen forceps are used. Omentum should be tacked over the incision to improve the seal and reduce the risk of leakage.

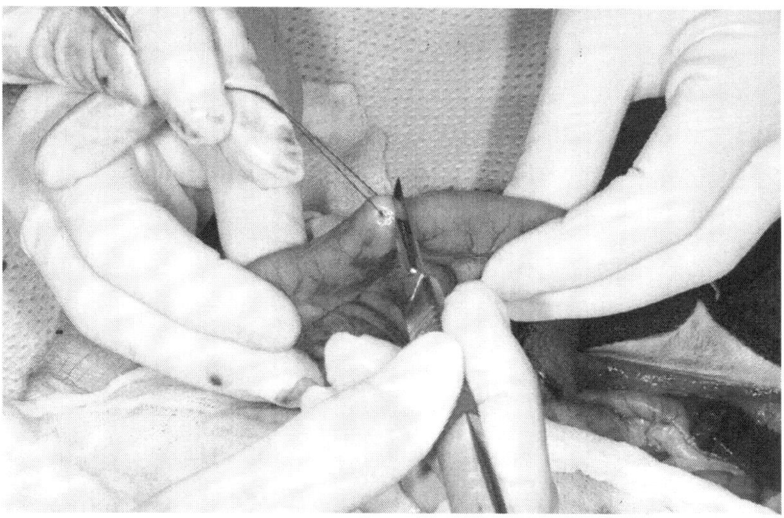

Fig. 7.7 Incisional biopsy of the small intestine.

Lymph Nodes

Lymph nodes that are enlarged or abnormal in texture can be sampled by aspiration, incisional biopsy, or excisional biopsy (Fig. 7.8). The lymph node should be carefully dissected free from surrounding mesentery or peritoneum, if possible. If complete excision is unsafe, a biopsy of a free end can be performed using a guillotine technique, or a piece of the lymph node can be removed with a wedge or shaving incision.

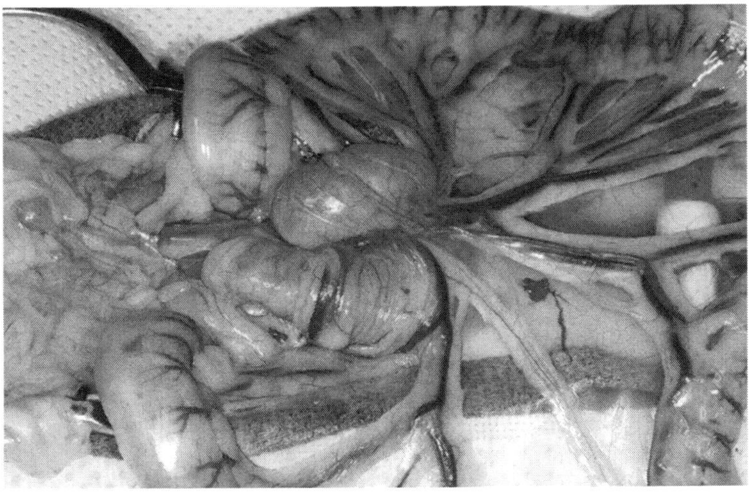

Fig. 7.8 Severely enlarged mesenteric lymph nodes.

PREPARING FOR CLOSURE

Before closure, the need for postoperative enteral nutrition and fluid support should be considered. Nasogastric tubes are easily placed and removed, and if the anesthetist advances the tube intraoperatively, the surgeon can confirm its placement within the stomach, obviating the need for postoperative radiographs.

If the gastrointestinal tract has been entered, gloves and instruments are changed. The abdomen can be lavaged with warm sterile saline to decrease inflammatory mediators, remove bacteria or contaminants, and warm the patient. If the procedure had minimal contamination, lavage can be limited to the area around the biopsy or incision site. If septic peritonitis is suspected, the peritoneum should be lavaged until the returning fluid is clear; this may require more than 200 mL/kg sterile saline. If significant contamination is still present after lavage, continuous suction drains can be placed for postoperative fluid removal.

Once all intraabdominal procedures are complete, sponges and laparotomy pads are counted to verify none have been left in the abdomen.

Anesthetic Considerations

The anesthetic management of animals with abdominal disease depends on the underlying disease. Animals that are not in shock can be premedicated with a benzodiazepine and opioid and induced with propofol, ketamine, or etomidate given intravenously to effect (Table 7.1). Table 7.2 provides suggested anesthetic protocols for animals that are in shock or that are debilitated.

Table 7.1: Anesthetic Considerations in the Stable Patient Undergoing Abdominal Surgery.

Preoperative Considerations	
Associated conditions	• Minimal; usually healthy
Blood work	• HCT • TP • In patients older than 5-7 yr, consider electrolytes, liver enzymes, BUN, and Cr
Physical examination	• May be painful if incarcerated hernia
Premedications	• If patient is anxious, give the following: ○ Diazepam (0.2 mg/kg, IV), ***or*** ○ Midazolam (0.2 mg/kg IV, IM), ***plus*** ○ Hydromorphone* (0.1-0.2 mg/kg IV, IM in dogs; 0.05-0.1 mg/kg IV, IM in cats), ***or*** ○ Morphine (0.1-0.2 mg/kg IV or 0.2-0.4 mg/kg IM), ***or*** ○ Buprenorphine† (0.005-0.02 mg/kg IV, IM)
Intraoperative Considerations	
Induction	• If premedicated, give: ○ Propofol (2-4 mg/kg IV) • If no premeds given, then: ○ Propofol (4-8 mg/kg IV), ***or*** ○ Ketamine (5.5 mg/kg IV) with diazepam (0.28 mg/kg IV), ***or*** ○ Dexmedetomidine‡ (33 µg/kg IM, plus butorphanol (0.66 mg/kg IM), plus ketamine (6.6 mg/kg IM); use one half of all doses if given IV
Maintenance	• Isoflurane or sevoflurane, ***plus*** ○ Fentanyl (2-10 µg/kg IV PRN in dogs; 1-4 µg/kg IV PRN in cats) for short-term pain relief, ***plus*** ○ Hydromorphone* (0.1-0.2 mg/kg IV PRN in dogs; 0.05-0.1 mg/kg IV PRN in cats), ***or*** ○ Morphine (0.1-1 mg/kg IV PRN in dogs; 0.05-0.2 mg/kg IV PRN in cats), ***or*** ○ Buprenoprhine† (0.005-0.02 mg/kg IV PRN), ***plus*** ○ Dexmedetomidine‡ (low-dose) (0.5-1 µg/kg IV), ***or*** ○ Medetomidine‡ (low-dose) (1-2 µg/kg IV), ***plus*** ○ Ketamine (low-dose) (0.5-1 mg/kg IV), ***or*** ○ Ketamine CRI (0.5 mg/kg IV loading dose, then 10 µg/kg/min IV)
Fluid needs	• 5-10 ml/kg/hr plus 3× EBL
Monitoring	• Blood pressure • HR • ECG • Respiratory rate • SpO_2 • Temperature
Blocks	Epidural: • Morphine (0.1mg/kg preservative free) ***or*** • Buprenorphine (0.003-0.005 mg/kg diluted in saline) Incisional: • Lidocaine (<5 mg/kg in dogs; 2-4 mg/kg in cats), ***or*** • Bupivicaine (<2 mg/kg)
Postoperative Considerations	
Analgesia	• Fentanyl CRI (1-10 µg/kg IV loading dose then 2-20 µg/kg/hr IV), ***or*** • Hydromorphone CRI (0.025-0.1µg/kg/hr IV in dogs), ***or*** • Morphine (0.1-1 mg/kg IV or 0.1-2 mg/kg IM q1-4hr in dogs; 0.05-0.2 mg/kg IV or 0.1-0.5 mg/kg IM q1-4hr in cats), ***or*** • Buprenorphine† (0.005-0.02 mg/kg IV, IM q4-8hr), ***or*** • Hydromorphone* (0.1-0.2 mg/kg IV, IM q3-4hr in dogs; 0.05-0.1 mg/kg IV, IM q3-4hr in cats), ***plus*** • +/– Ketamine CRI (2 µg/kg/min IV. If no previous loading dose, give 0.5 mg/kg IV prior to CRI), ***plus*** • In dogs: ○ Carprofen (2.2 mg/kg q12hr PO), ***or*** ○ Deracoxib (3-4 mg/kg q24hr for <7 days PO), ***or*** ○ Meloxicam (0.1-0.2 mg/kg SC, PO, once then 0.1 mg/kg PO q24hr) • In cats: ○ Meloxicam§ (0.05-0.1 mg/kg PO or SC once), ***or*** ○ Buprenorphine† (0.01-0.02 mg/kg OTM q6-12hr)
Monitoring	• SpO_2 • Blood pressure • HR • Respiratory rate • Temperature
Blood work	HCT and TP if large blood loss
Estimated pain score	Can be severe depending on source of pain and/or procedure. Some of these patients have been in chronic pain.

BUN, Blood urea nitrogen; Cr, creatinine; CRI, constant rate infusion; EBL, estimated blood loss; HCT, hematocrit; HR, heart rate; SpO2, oxygen saturation via a pulse oximeter; TP, total protein; PRN, as needed; OTM, oral transmucossal.

*Monitor for hyperthermia in cats.

†Buprenorphine is a better analgesic than morphine in cats.

‡Only use in young, healthy animals.

§Black box warning added by the FDA in October 2010 identified cases of renal failure and death in cats with repeated uses of meloxicam. Meloxicam is approved for single use only in cats in the U.S.

Table 7.2: Anesthetic Considerations in the Septic Patient.

Preoperative Considerations	
Associated conditions	• Dehydration • Electrolyte abnormalities • Hypotension • Abnormal blood glucose • Anemia • Adrenal suppression may be present in the critically ill patient
Blood work	• HCT • Electrolytes • BUN • Cr • TP • Lactate • Blood glucose, often serial glucose checks • Urinalysis
Physical examination	• Often a younger patient that was previously healthy • May be dehydrated, tachycardic or bradycardic, hypotensive, vomiting, and/or hypothermic • Painful abdomen may be present
Other diagnostics	• Blood pressure • ECG
Premedications	• Rehydrate over 4-6 hours if possible; if emergent, may have to give more rapid boluses to expedite time to surgery • Correct electrolyte and glucose abnormalities. • Avoid sedatives in depressed patients • Avoid alpha 2 agonists and acepromazine. • If patient is anxious, give the following: ○ Midazolam (0.1-0.2 mg/kg IV, IM) ***or*** ○ Diazepam (0.1-0.2 mg/kg IV) • If patient is not depressed, then give the following: ○ Hydromorphone* (0.1-0.2 mg/kg IV, IM in dogs; 0.05-0.1 mg/kg IV, IM in cats), ***or*** ○ Morphine (0.1-0.2 mg/kg IV or 0.2-0.4 mg/kg IM) ***or*** ○ Buprenorphine† (0.005-0.02 mg/kg IV, IM)

Continued

Intraoperative Considerations	
Induction	• If dehydrated, give the following: ○ Etomidate (0.5-1.5 mg/kg IV); if possible, avoid its use in critically ill patients, ***or*** ○ Propofol (1-4 mg/kg IV slowly), ***or*** • If hydrated, give the following: ○ Propofol (2-4 mg/kg IV if sedated or 4-8 mg/kg IV if unsedated)
Maintenance	• Isoflurane or sevoflurane, ***plus*** ○ Fentanyl (2-10 µg/kg IV PRN in dogs; 1-4 µg/kg IV PRN in cats) for short-term pain relief, ***plus*** ○ Fentanyl CRI (1-5 µg/kg IV loading dose, then 2-30 µg/kg/hr IV), ***or*** ○ Hydromorphone* (0.1-0.2 mg/kg IV PRN in dogs; 0.05-0.1 mg/kg IV PRN in cats), ***or*** ○ Buprenorphine† (0.005-0.02 mg/kg IV PRN), ***plus*** PRN ○ Ketamine (low-dose) (0.5-1 mg/kg IV), ***or*** ○ Ketamine CRI (0.5 mg/kg IV loading dose, then 10 µg/kg/min IV) • For hypotension (to keep MAP 60-80 mm Hg), give phenylephrine, ephedrine, norepinephrine, or dopamine as needed
Fluid needs	• 10 ml/kg/hr if minimal EBL and minimal evaporative losses, or 10-20 ml/kg/hr if open abdomen with higher evaporative losses, plus 3x EBL. Higher rates of fluids are necessary if dehydration not corrected preoperatively • Consider colloids if persistent hypotension
Monitoring	• Blood pressure • ECG • Respiratory rate • SpO_2 • $EtCO_2$ • Temperature • U/O
Blocks	Epidural: • Morphine (0.1 mg/kg preservative free) ***or*** • Buprenorphine (0.003-0.005 mg/kg diluted in saline) • Avoid local anesthetics for spinals and epidurals in hypotensive patients. Incisional: • Lidocaine (< 5 mg/kg in dogs; 2-4 mg/kg in cats), ***or*** • Bupivicaine (< 2 mg/kg)
Postoperative Considerations	
Analgesia	• Fentanyl CRI (1-10 µg/kg IV loading dose, then 2-20 µg/kg/hr IV), ***or*** • Morphine (0.1-1 mg/kg IV or 0.1-2 mg/kg IM q1-4hr in dogs; 0.05-0.2 mg/kg IV or 0.1-0.5 mg/kg IM q1-4hr in cats) if not hypotensive, ***or*** • Hydromorphone* (0.1-0.2 mg/kg IV, IM q3-4hr in dogs; 0.05-0.1 mg/kg IV, IM q3-4hr in cats) ***or*** • Hydromorphone CRI (0.025-0.1 mg/kg/hr IV in dogs), ***or*** • Buprenorphine† (0.005-0.02 mg/kg IV, IM q4-8hr; 0.01-0.02 mg/kg OTM q6-12hr in cats), ***plus*** • ±Ketamine CRI (2 µg/kg/min IV. If no previous loading dose, give 0.5 mg/kg IV prior to CRI) • Avoid NSAIDs in hypotensive patients.
Monitoring	• SpO_2 • Blood pressure • HR • Respiratory rate • Temperature • U/O • ECG if electrolyte abnormalities
Blood work	• HCT if significant blood loss • Repeat abnormal preoperative blood work • Serial blood glucose checks if necessary
Estimated pain score	Moderate to severe if open abdominal surgery or if underlying pancreatitis

BUN, Blood urea nitrogen; Cr, creatinine; CRI, constant rate infusion; EBL, estimated blood loss; ECG, electrocardiogram; EtCO2, end-tidal CO2; HCT, hematocrit; HR, heart rate; MAP, mean arterial pressure; SpO2, oxygen saturation via a pulse oximeter; TP, total protein; U/O, urine output; PRN, as needed; OTM, oral transmucosal.

Surgical Anatomy

The rectus sheath is composed of an external and an internal leaf. The external leaf is formed by the aponeurosis of the external abdominal oblique muscle and a portion of the aponeurosis of the internal abdominal oblique muscle. The aponeurosis of the transversus abdominis muscle joins the external leaf near the pubis. The internal leaf consists of a portion of the aponeurosis of the internal abdominal oblique muscle, the aponeurosis of the transversus abdominis muscle, and the transversalis fascia. The internal leaf disappears in the caudal third of the abdomen, where the aponeurosis of the internal abdominal oblique muscle joins the external leaf, leaving the caudal rectus abdominis muscle covered only by a thin sheet of transversalis fascia and peritoneum.

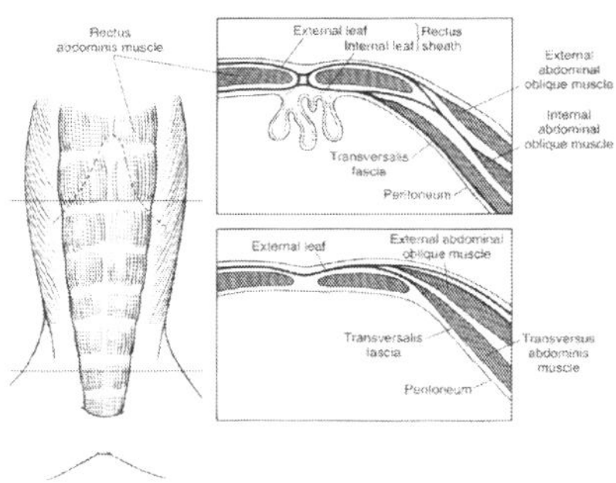

Fig. 7.9 Anatomy of the rectus sheath.

NOTE • The linea alba is easier to locate near the umbilicus because it becomes thinner near the pubis.

Surgical Techniques

The abdomen generally is explored by means of a ventral midline incision. In most animals, the entire abdomen, including the inguinal areas and the caudal thorax, should be prepared for aseptic surgery to allow extension of the incision into thoracic or pelvic cavities if necessary. Prepping too small an area is a common mistake, particularly for abdominal exploration in trauma patients. To visualize all abdominal structures adequately, the incision must extend from the xiphoid process to the pubis. If only a specific abdominal structure will be examined, a shorter incision can be made. A caudal abdominal incision extending from umbilicus to pubis is adequate for bladder exploration; similarly, a cranial abdominal incision (i.e., umbilicus to xiphoid process) allows evaluation of liver and stomach. Occasionally, the midline incision is extended laterally at the xiphoid process (1 cm caudal to the last rib) to facilitate exposure of the liver, biliary system, and diaphragm. A paracostal (paralumbar) celiotomy can be used to expose the kidneys and adrenal glands; it is most commonly used for unilateral adrenalectomy.

VENTRAL MIDLINE CELIOTOMY IN CATS AND FEMALE DOGS

With the patient in dorsal recumbency, make a ventral midline skin incision beginning near the xiphoid process and extending caudally to the pubis. Sharply incise the subcutaneous tissues until the external fascia of the rectus abdominis muscle is exposed. Ligate or cauterize small subcutaneous bleeders and identify the linea alba. Tent the abdominal wall and make a sharp incision into the linea alba with a scalpel blade. Palpate the interior surface of the linea for adhesions. Use scissors to extend the incision cranially or caudally (or both) to near the extent of the skin incision. Digitally break down the attachments of one side of the falciform ligament to the body wall, or excise it and remove it entirely if it interferes with visualization of cranial abdominal structures. Clamp the cranial end of the falciform ligament and ligate or cauterize bleeders before removing it.

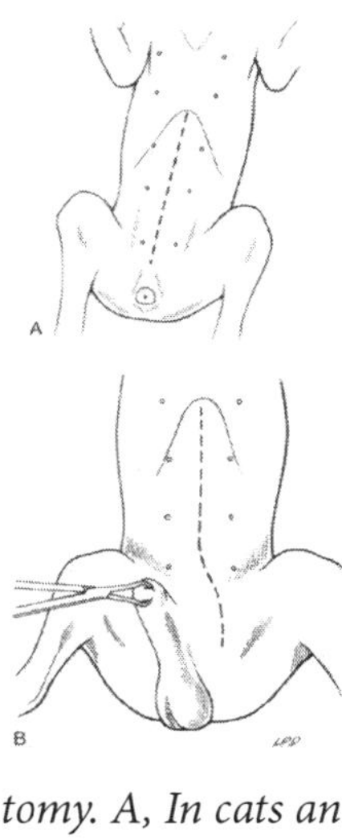

Fig. 7.10 Ventral midline celiotomy. A, In cats and female dogs. B, In male dogs.

Ventral Midline Celiotomy in Male Dogs

With the patient in dorsal recumbency, place a towel clamp on the prepuce and clamp it to the skin on one side of the body. Drape the tip of the prepuce and clamp outside the surgical field. Make a ventral midline skin incision beginning at the xiphoid process and continuing caudally to the prepuce. Curve the incision to the left or right of the penis and prepuce (i.e., the side opposite the clamped prepuce) and extend it to the level of the pubis. Incise the subcutaneous tissues and fibers of the preputialis muscle to the level of the rectus fascia in the same plane as the skin incision. Ligate or cauterize large branches of the caudal superficial epigastric vein at the cranial aspect of the prepuce. Retract incised skin and subcutaneous tissues laterally and locate the linea alba and external fascia of the rectus abdominis muscle. Do not attempt to locate the caudal linea alba until subcutaneous tissues have been incised and the abdominal musculature fascia identified. Tent the abdominal wall and make a sharp incision into the linea alba with a scalpel blade. Palpate the interior surface of the linea for adhesions. Use scissors to extend the incision cranially or caudally (or both) to near the extent of the skin incision.

Paracostal Celiotomy

Position the animal in lateral recumbency and place a rolled towel or sandbag between the animal and the operating table. Make a skin incision from the ventral vertebral column to near the ventral midline. Center the incision halfway between the wing of the ilium and the last rib. Extend the incision through the external abdominal oblique muscle with scissors. Separate internal abdominal oblique and transversus abdominis muscle fibers and expose the peritoneal and transversalis fascia. Tent the peritoneum and sharply incise it with scissors.

Abdominal Exploration

Systematically explore the entire abdomen. Various techniques may be used; however, every surgeon should develop a consistent pattern to ensure that the entire abdominal cavity and all structures in each animal are visualized and/or palpated.

1. Explore the cranial quadrant.

- Examine the diaphragm (including the esophageal hiatus) and the entire liver (palpate the liver).
- Inspect the gallbladder and biliary tree; express the gallbladder to determine its patency.
- Examine the stomach, pylorus, proximal duodenum, and spleen.
- Examine both pancreatic limbs (palpate gently!), the portal vein, hepatic arteries, and caudal vena cava.

2. Explore the caudal quadrant.

- Inspect the descending colon, urinary bladder, urethra, and prostate or uterine horns.
- Inspect the inguinal rings.

3. Explore the intestinal tract.

- Palpate the intestinal tract from the duodenum to the descending colon and observe the mesenteric vasculature and nodes.

4. Explore the gutters.

- Use the mesoduodenum to retract the intestine to the left, and examine the right "gutter." Palpate the kidney, and examine the adrenal gland, ureter, and ovary.
- Use the descending colon to retract the abdominal contents to the right. Examine the left kidney, adrenal gland, ureter, and ovary.

Use moistened laparotomy sponges to protect tissues from drying during the procedure. If generalized infection is present, or if diffuse intraoperative contamination has occurred, flush the abdomen with copious amounts of warmed, sterile saline solution.

Historically, many different antiseptics (e.g., povidone-iodine, chlorhexidine) and antibiotics have been added to lavage fluids. Povidone-iodine is the most widely used antiseptic; however, this practice has not shown a beneficial effect in repeated experimental and clinical trials and may be detrimental in animals with established peritonitis because the carrier, polyvinylpyrrolidone, inhibits macrophage chemotaxis. Similarly, no substantial evidence indicates that adding antibiotics to lavage fluid benefits patients treated with appropriate systemic antibiotics. Room temperature lavage fluids should not be used in anesthetized patients. Heated lavage fluids are useful for increasing the temperature in dogs.

Remove the lavage fluid and blood and inspect the abdominal cavity before closure to ensure that all foreign material and surgical equipment have been removed. Perform a sponge count and compare it with the preoperative count to ensure that surgical sponges have not been left in the abdominal cavity.

ABDOMINAL WALL CLOSURE

The linea alba may be closed with simple interrupted sutures or a simple continuous suture pattern. The simple continuous technique does not increase the risk of dehiscence when properly performed (i.e., secure knots, appropriate suture material), and it allows for rapid closure. Preferably strong, absorbable suture material (e.g., polydioxanone [PDS], polyglyconate [Maxon], poliglecaprone [Monocryl], glycomer 631 [Biosyn]) should be used for continuous suture patterns, and six to eight knots should be placed at each end of the incision line. Monofilament, nonabsorbable suture material (e.g., polybutester [Novafil], polypropylene [Prolene], nylon) has been associated with suture sinus formation and should be avoided. Surgical gut and stainless steel wire should not be used for continuous suture patterns.

On each side of the incision, incorporate 4 to 10 mm of fascia in each suture. Place interrupted sutures 5 to 10 mm apart, depending on the animal's size. Tighten sutures sufficiently to appose but not enough to strangulate tissue, because too-tight sutures adversely affect wound healing. Incorporate full-thickness bites of the abdominal wall in the sutures if the incision is midline. Do not incorporate the falciform ligament between the fascial edges. If the incision is lateral to the linea alba and muscular tissue is exposed (i.e., paramedian incision), close the external rectus sheath without including muscle in the sutures. Do not attempt to include peritoneum in the sutures. Close subcutaneous tissues with a simple continuous pattern of absorbable suture material, and reappose the preputialis muscle fibers. Use nonabsorbable sutures or stainless steel staples to close the skin. Place skin sutures without tension.

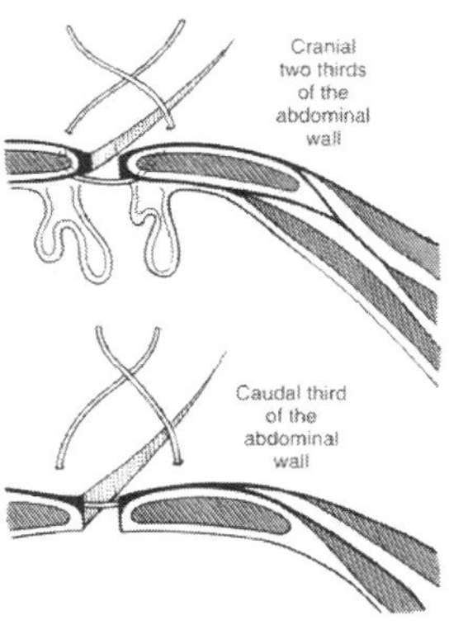

Fig. 7.11 To close a midline incision, incorporate full-thickness bites of the linea alba (or the external sheath only) in the sutures.

For paracostal celiotomy, close the individual muscle layers with synthetic absorbable suture material in a continuous or interrupted pattern. Attempt to eliminate dead space between muscle layers. Appose subcutaneous tissue with absorbable suture in a continuous or interrupted pattern, and close the skin with nonabsorbable suture in a simple interrupted or continuous pattern.

HEALING AND THE ABDOMINAL WALL

The ability of tissues to hold sutures without tearing depends on the strength of the tissue and the orientation of collagen fibrils. Skin and fascia are strong, whereas muscle and fat are weak. Peritoneum heals rapidly across the incision and does not contribute to wound strength; therefore, closure of this layer is not beneficial. Experimental and clinical studies in dogs suggest that suturing peritoneum may increase the incidence of postoperative intraabdominal adhesions.

Suture Materials And Special Instruments

Useful instruments for celiotomy include Balfour abdominal retractors, Poole or Yankauer suction tips, malleable retractors, and Mixter (right-angle) forceps. Laparotomy pads and 4 × 4 sponges should have radio-opaque markers. See the previous discussion on abdominal wall closure for choice of suture material.

Postoperative Care And Assessment

The abdominal incision should be checked twice daily for redness, swelling, or discharge. If the animal licks or chews at the incision, an Elizabethan collar or sidebar should be used to prevent iatrogenic suture removal. Early signs of altered wound healing include inflammation and edema. Swelling and serosanguineous drainage from the incision are consistent signs of acute incisional dehiscence. Dehiscence usually occurs 3 to 5 days after surgery, when minimal healing has occurred and the sutures have weakened; however, it may occur earlier if knots were tied improperly or if fascia was not incorporated into the sutures. Evisceration usually causes sepsis and severe blood loss secondary to mutilation of exposed intestine; the patient must be treated promptly. The abdomen should be bandaged, fluid therapy initiated, and broad-spectrum antibiotics given while the animal is prepared for surgery. If technical failure such as poor knot tying or improper suturing is suspected, the entire suture line should be removed and replaced. Débridement of the wound edges is unnecessary and delays wound healing. The intestine should be closely inspected for viability and damaged sections resected if appropriate (see p. 501). The abdominal cavity should be lavaged copiously with warmed, sterile saline. Open abdominal drainage (see p. 381) or suction drainage may be considered in animals with generalized peritonitis. Wound disruption after 10 to 21 days usually causes hernia formation rather than evisceration. Hernia repair in these animals may require excision of fibrotic tissues. Subsequent closure requires that tissue layers be accurately apposed.

UMBILICAL AND ABDOMINAL HERNIAS

Definitions

External abdominal hernias are defects in the external wall of the abdomen that allow protrusion of abdominal contents; internal abdominal hernias are those that occur through a ring of tissue confined within the abdomen or thorax (e.g., diaphragmatic hernia, hiatal hernia). External abdominal hernias may involve the abdominal wall anywhere other than the umbilicus, inguinal ring, femoral canal, or scrotum. Umbilical hernias occur through the umbilical ring. The contents of true hernias generally are enclosed in a peritoneal sac; false hernias allow protrusion of organs outside a normal abdominal opening; therefore the contents seldom are contained in a peritoneal sac. Omphaloceles are large midline umbilical and skin defects.

Abdominal hernias may be defined according to their location (i.e., ventral, prepubic, subcostal, hypochondral, paracostal, or lateral). The cranial pubic ligament formerly was called the prepubic tendon.

Abdominal hernias generally occur secondary to trauma, such as vehicular accidents or bite wounds; however, they occasionally occur as congenital lesions. Congenital cranial abdominal hernias (i.e., cranial to the umbilicus) have been reported in association with peritoneopericardial diaphragmatic hernias in dogs and cats. Abdominal hernias are false hernias because they do not contain a hernial sac. When associated with blunt trauma, they arise as a result of rupture of the wall from within caused by an increase in intraabdominal pressure while the abdominal muscles are contracted. The most common sites of traumatic abdominal hernias are the prepubic region and the flank. Cranial pubic ligament hernias often occur in association with pubic fractures. Paracostal hernias may allow migration of abdominal contents along the thoracic wall. In rare cases, the abdominal contents enter the chest through defects in the intercostal muscles. Nearly half of the animals that suffer from traumatic abdominal hernia have serious concurrent injuries, including orthopedic (e.g., pelvis) and soft tissue injuries; thus a thorough physical examination of all affected animals should be performed.

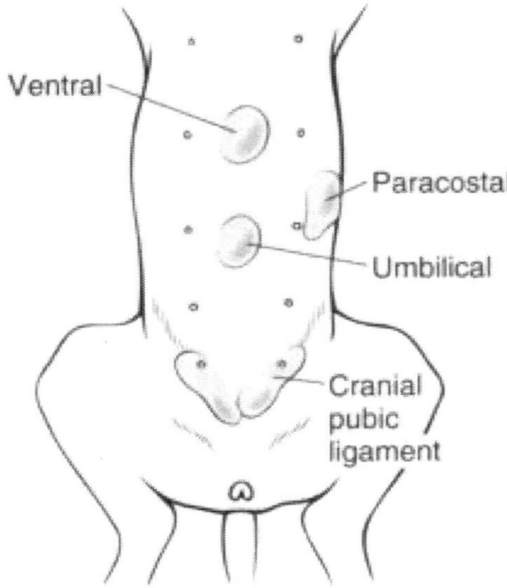

Fig. 7.12 Location of abdominal and umbilical hernias.

Umbilical hernias usually are congenital, caused by flawed embryogenesis. Umbilical vessels, the vitelline duct, and the stalk of the allantois pass through the umbilical ring in the fetus, but this aperture closes at birth, leaving an umbilical cicatrix. If the aperture fails to contract or is too large or is improperly formed, a hernia results. These hernias are lined by a peritoneal sac and are considered true hernias. The cause of umbilical hernias is seldom known, but most are thought to be inherited. Many male dogs with umbilical hernias are cryptorchid. Omphaloceles allow abdominal organs to protrude externally (eviscerate). The abdominal contents initially are covered by amniotic tissue, but this membrane covering is easily ruptured. Most affected neonates die or are euthanized at birth.

Diagnosis

Clinical Presentation

Signalment

Most animals with umbilical or abdominal hernias are young. Umbilical hernias are believed to be heritable in some breeds (e.g., Airedale, Basenji, Pekingese). Cranial ventral abdominal hernias associated with peritoneopericardial diaphragmatic hernias may be inherited in Weimaraners.

History

A history of trauma is common with abdominal hernias. The hernia initially may be overlooked while more obvious or life-threatening injuries are treated. Small umbilical hernias often are not noticed until the animal is examined for neutering. If strangulation or intestinal obstruction occurs, the animal may be presented for treatment of vomiting, abdominal pain, anorexia, and/or depression.

PHYSICAL EXAMINATION FINDINGS

Abdominal structures (e.g., organs, omentum) in the subcutaneous space or between muscle layers usually cause asymmetry of the abdominal contour. The size of the swelling may not correspond to the size of the hernia, particularly if intestine has migrated into the hernia. The swelling should be palpated carefully to discern the contents of the hernia (e.g., intestine, bladder, spleen) and to locate the abdominal defect. These patients should be thoroughly examined to determine whether a concurrent abdominal or thoracic injury or abnormality exists. Rupture of the cranial pubic ligament often is difficult to palpate because of subcutaneous swelling and pain.

Umbilical hernias usually manifest as a soft ventral abdominal mass at the umbilical scar. Deep palpation of the swelling reveals the size of the umbilical ring and helps characterize hernial contents. The hernial ring is not palpable in some animals because the ring closes subsequent to herniation of falciform fat or omentum. Occasionally, intestine or other abdominal structures can be palpated; they generally can be reduced into the abdominal cavity. If the umbilical sac is warm or painful and the contents are irreducible, intestinal strangulation or obstruction should be suspected.

Diagnostic Imaging

Radiographs should be taken in animals with abdominal hernias. Routine ventral dorsal and lateral views may show an associated abdominal or thoracic injury (e.g., abdominal fluid, diaphragmatic hernia). Abdominal radiographs may help confirm a hernia (i.e., subcutaneous intestinal loops and loss of the ventral abdominal stripe) when the abdominal wall defect cannot be palpated because of swelling or pain. Radiographs generally are not indicated in small umbilical hernias. Ultrasound examination may help define the contents of hernias.

Laboratory Findings

Laboratory abnormalities are uncommon with umbilical hernias unless strangulation or intestinal obstruction is present. Abnormalities associated with abdominal hernias vary depending on the severity of concurrent internal injuries.

Differential Diagnosis

Most hernias are diagnosed on physical examination. Differential diagnoses for abdominal swellings include abscess, cellulitis, hematoma or seroma, and neoplasia.

Medical Management

Initial treatment of animals with an abdominal hernia is directed toward diagnosing and treating shock and concurrent life-threatening internal injuries.

Surgical Treatment

Most abdominal hernias can be repaired by suturing torn muscle edges or apposing a disrupted abdominal wall edge to the pubis, ribs, or adjacent fascia. Synthetic mesh must be used to repair the defect in rare cases. Some hernias (e.g., intestinal strangulation, urinary obstruction, concurrent organ trauma) require emergency surgical correction. However, the extent of devitalized muscle may not be apparent initially, and delaying surgery in stable patients until muscle damage can be accurately assessed facilitates surgical correction. The most common surgical complications are hernia recurrence and wound infection. Abdominal hernias secondary to bite wounds usually are contaminated; wound infection and dehiscence of the skin or hernial repair (or both) are common. Nonabsorbable mesh should not be placed in these hernias, and the wounds should be drained. Treatment of infected wounds includes cultures, drainage, antibiotics, and/or flushing. Abdominal exploration should be performed at herniorrhaphy to diagnose concurrent abdominal organ injury (e.g., mesenteric avulsion, gastric or intestinal perforation, diaphragmatic herniation, bladder rupture).

Many umbilical hernias resolve spontaneously in young animals or are small and are not corrected until the animal is neutered. Spontaneous closure may occur as late as 6 months of age. Intestinal strangulation is most likely to occur when the hernial defect is about the size of the intestine and the hernial sac is large. Strangulation is unlikely with very small or large defects. If abdominal viscera in the hernia cannot be reduced, surgery should be performed as soon as possible.

PREOPERATIVE MANAGEMENT

Preoperative care depends on the animal's status and concurrent injuries. Hydration and electrolyte abnormalities should be corrected before surgery.

Anesthesia

If no concurrent abdominal injury or disease is identified, a variety of anesthetic protocols can be used to anesthetize the animal. Anesthetic management of animals that are in shock or are debilitated. See also subsequent chapters for detailed information about anesthetic management of patients with specific diseases (e.g., renal, pancreatic, hepatic).

Surgical Anatomy

The abdominal wall is composed of four muscle layers (the external and internal abdominal oblique muscles, the rectus abdominis muscle, and the transversus abdominis muscle).

Abdominal hernias may occur at insertions or attachments of these muscles or through muscle bellies themselves. In the dog, the cranial pubic ligament (prepubic tendon) is a band of transverse fibers that serves to attach the ventral abdominal muscles to the cranial border of the pubis. In cats, no distinct prepubic tendon exists (Beittenmiller et al, 2009); rather, the abdominal muscles attach directly to the pelvic brim.

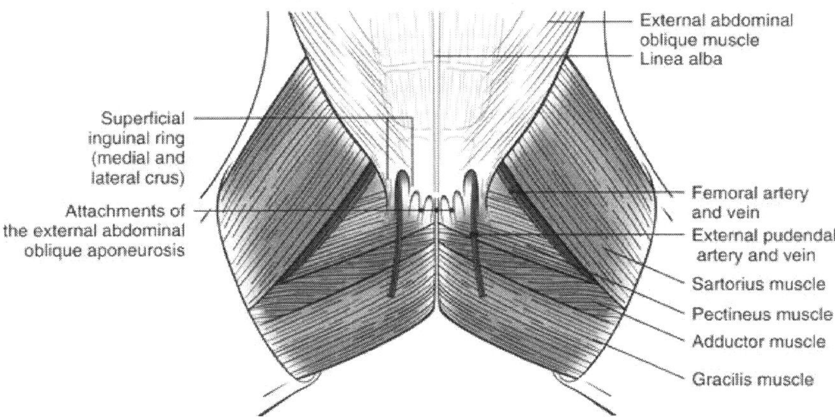

Fig. 7.13 Normal anatomy of the prepubic musculature attachments in the cat. (Modified from Beittenmiller MR, Mann FA, Constantinescu GM, et al, Clinical anatomy and surgical repair of prepubic hernia in dogs and cats, J Am Anim Hosp Assoc 45:284, 2009.)

Positioning

For ventral hernias, the animal is placed in dorsal recumbency and the area around the hernia is prepared for aseptic surgery. Repair of rupture of the cranial pubic ligament may be facilitated by placing the animal in dorsal recumbency with the rear limbs flexed and pulled cranially.

Surgical Techniques

Abdominal Hernias

For most abdominal hernias, perform a ventral midline abdominal incision to allow the entire abdomen to be explored. Assess the extent of visceral herniation. Reduce the herniated contents and amputate or excise necrotic or devitalized tissue around the hernia. Close the muscle layers of the hernia with simple interrupted or simple continuous sutures. If a large area of devitalized tissue is removed, use synthetic mesh such as Marlex or Prolene to close the defect (do not place nonabsorbable mesh in infected sites). Fold the edges of the mesh over and suture the folded edges to viable tissue using simple interrupted sutures. Injuries to the cranial pubic ligament can be difficult to repair. If necessary, drill holes in the pubic bone to anchor the sutures.

PARACOSTAL HERNIAS

Make a midline incision or make one directly over the hernia. Explore the hernia and suture the torn edges of the transverse, internal, and external abdominal oblique muscles. Incorporate a rib in the suture if muscle has been avulsed from the costal arch.

Cranial pubic ligament (prepubic) hernias

In dogs, make a ventral midline skin incision and identify the ruptured tendon and its pubic insertion. Evaluate the inguinal rings and vascular lacuna; these hernias may extend into the femoral region as a result of rupture of the inguinal ligament. In cats, inspect the crura of the superficial inguinal ring and the aponeurosis of the external abdominal oblique muscle for damage. Reattach the free edge of the abdominal wall to the cranial pubic ligament with simple interrupted, cruciate, or mattress sutures of nonabsorbable monofilament suture (e.g., polybutester [Novafil], polypropylene [Prolene], nylon) or absorbable suture material (see later under "Suture Materials and Special Instruments"). If an inguinal or femoral hernia is present, reattach the prepubic tendon to the pubis first, before repairing the other defects.

As an alternative, suture the tendon remnant to the muscle fascia and periosteum covering the pubis, or anchor it to the pubis by drilling holes in the pubic bone through which sutures can be placed. If the hernia extends into the femoral region, it may be necessary to suture the body wall to the medial fascia of the adductor muscles. When doing so, take care to avoid damaging the femoral vessels or nerves.

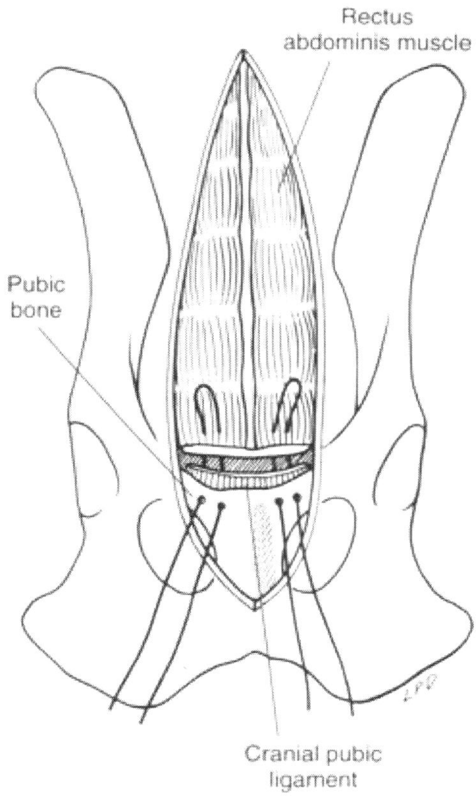

Fig. 7.14 To repair injury to the cranial pubic ligament, it may be necessary to anchor the ligament to the pubis by drilling holes in the pubic bone through which sutures can be placed.

A cranial sartorius muscle flap may be used to repair prepubic tendon injuries. Separate the cranial portion of the muscle from its caudal belly and the quadriceps femoris muscle. Then transect the cranial portion of the muscle at its distal insertion on the medial aspect of the patella and elevate it to the level of the proximal vascular pedicle. Rotate the flap 180 degrees to the defect, and secure it using simple interrupted sutures. Use the flap to repair concurrent inguinal or femoral hernia, if present.

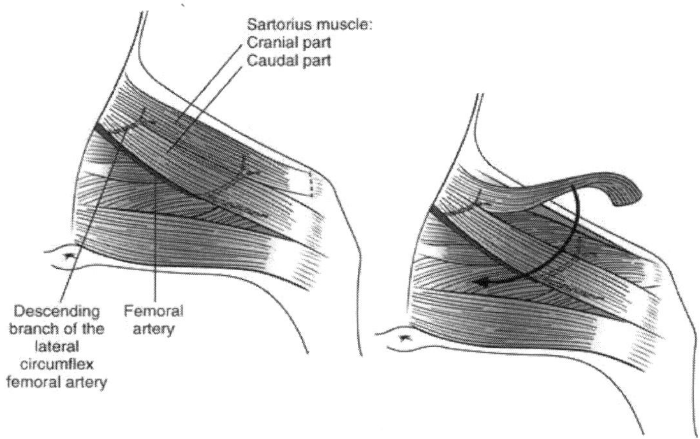

Fig. 7.15 Sartorius muscle flap in a cat. Transect the sartorius muscle distally at its insertion on the medial aspect of the patella, and elevate it to the level of the proximal vascular pedicle. (Modified from Beittenmiller MR, Mann FA, Constantinescu GM, et al, Clinical anatomy and surgical repair of prepubic hernia in dogs and cats, J Am Anim Hosp Assoc 45:284, 2009.)

Umbilical Hernias

For umbilical hernias, palpate the hernial ring, reduce the abdominal contents if possible, and incise the skin over the umbilicus. If the hernia contains only fat or omentum, ligate the hernial neck and excise the sac and its contents. As an alternative, if adhesions are not present, invert the sac and its contents into the abdominal cavity. Do not débride the wound margins. Suture the edges of the defect with monofilament, synthetic, absorbable suture (e.g., polydioxanone [PDS], polyglyconate [Maxon], poliglecaprone 25 [Monocryl], glycomer 631 [Biosyn]) in a simple interrupted pattern. If the hernial contents cannot be reduced, make an elliptic incision around the swelling to avoid damaging the contents. Incise the hernial sac and replace the contents in the abdominal cavity. If the contents are irreducible, or if strangulation or intestinal obstruction is present, extend the abdominal defect on the midline. Explore the abdomen and inspect the intestines for viability before closing the defect. Umbilical hernia repair seldom requires mesh implantation.

SUTURE MATERIALS AND SPECIAL INSTRUMENTS

Strong, absorbable suture (e.g., polydioxanone [PDS], polyglyconate [Maxon], poliglecaprone 25 [Monocryl], glycomer 631 [Biosyn]) or nonabsorbable suture (e.g., polybutester [Novafil], polypropylene [Prolene], nylon) should be used to repair abdominal or ventral hernias. Nonabsorbable suture materials may be preferred over absorbable suture materials because they maintain tensile strength for longer than a year (see p. 68); however, successful repair with absorbable suture materials has been reported. Marlex and Prolene synthetic mesh may be used to repair some large defects. A study of polypropylene mesh showed a substantial reduction in the diameter of the ductus deferens associated with an intense inflammatory reaction when the mesh was placed in contact with the spermatic cord of dogs (Goldenberg and Ferreira de Paula, 2005).

Postoperative Care And Assessment

Postoperative care of these patients is dictated by the presence of concurrent injury or disease. The patient should be kept quiet, and the wound should be checked frequently for infection or dehiscence. Vomiting, fever, and/or leukocytosis may indicate peritonitis (see p. 374).

Prognosis

The prognosis generally is good, and recurrence is uncommon. When recurrence occurs, it generally is noted within a few days of surgery. Most animals have excellent long-term results when appropriate techniques are used.

TIPS TO REMEMBER WHEN PERFORMING ABDOMINAL SURGERY:

- Do not grab skin with forceps!
- Use the pencil grip on the thumb forceps (they are not tweezers or barbecue tongs).
- Do not dissect too wide creating lots of dead space but expose the linea alba well enough to identify where to enter and to help ensure you include the external rectus fascia when closing the abdominal cavity.
- Do not pull long suture strand too far away from your surgical site. Instead, re-grasp the suture material closer to the surgical field to prevent contamination (on the light or surgical mask). Keep instruments, hands and suture within the surgical field / sterile area at all times.
- When tying knots, keep the suture ends loose, watch the knot form, keep your hands parallel (flat) to prevent forming slip knots. Tie your knots slowly to ensure good apposition (don't rush).
- Do not manipulate / grasp the suture strand with needle holders or other instruments other than at the tip / end of the suture that will be cut / removed. Instruments damage the suture strand.

- Follow the curve of the needle when passing through tissues to prevent excessive tissue trauma. Grasp the suture needle with instruments (not fingers) when pulling it out of the tissues.
- Do not use surgeon throws unless there is tension on the tissues or a lot of fat on the pedicle. Include the external rectus fascia in all bites when closing the abdominal wall. It is the layer of strength.
- Take bites close to the skin when closing the subcutaneous tissues in order to appose the skin fairly closely. This decreases tension on the skin sutures and allows placement of loose skin sutures to accommodate postoperative swelling. Tack down to the external rectus fascia every 2-3 bites in order to decrease dead space and potential seroma formation. Do not include or damage the linea sutures with your needle.
- Burrying the knot is considered essential and should be performed in the subcutaneous tissues and for intradermal / subcuticular closure. If the knot doesn't burry, consider placing a single skin suture overtop and removing it before discharge. Do not use tissue glue.
- Replace instruments and gauzes on the instrument table after use and use a single gauze at a time to prevent loosing any in the surgical field.
- Remember to ensure the suture you are using is what you think it is. Keep non-absorbable suture meant for skin closure at the back of the table so you don't inadvertently use it within the abdominal cavity.
- Intradermal closure must be performed using cutting needle.

ABDOMINAL EXPLORATORY: A THERAPEUTIC AND DIAGNOSTIC TOOL FOR VETERINARIANS

Abdominal exploratory surgery—a systematic evaluation of the abdominal cavity and its contents—is often done in veterinary practice for therapeutic reasons, including elective sterilization, retrieval of urinary calculi, gastrointestinal foreign body removal and treatment of neoplastic conditions and other abdominal maladies. Looking beyond the therapeutic benefits, abdominal exploratory is an invaluable diagnostic tool as well, particularly in patients with nonspecific chronic gastrointestinal signs, hepatopathy or abnormal findings on radiographic or ultrasonographic examination. Acquiring good biopsy samples is key to maximizing the information gained during an exploratory.

An exploratory can be done via open laparotomy or laparoscopy. Open laparotomy has the advantage of allowing visual and tactile inspection as well as the ability to address any problems identified. Laparoscopy is a minimally invasive alternative that allows good visual inspection of the abdomen, acquisition of excellent biopsy samples and completion of many procedures.

Steps for successful abdominal exploratory

1. Prepare the patient and operating room. A complete sponge count (4-x-4 gauze sponges and laparotomy sponges) and instrument count (hemostats) should be done before entering any body cavity.

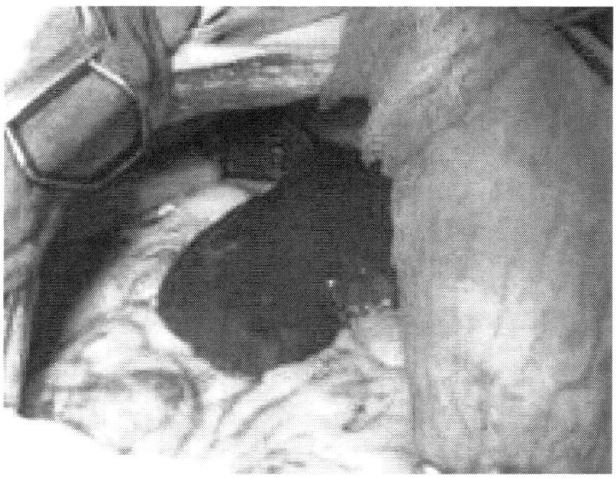

Fig. 7.16 An abdominal exploratory revealing splenic neoplasia.

2. Approach the abdomen via the ventral midline through the linea alba. A full exploratory is best done with an incision from the xyphoid to pubis. Excision of the falciform ligament and associated fat allows better access to the cranial abdomen.

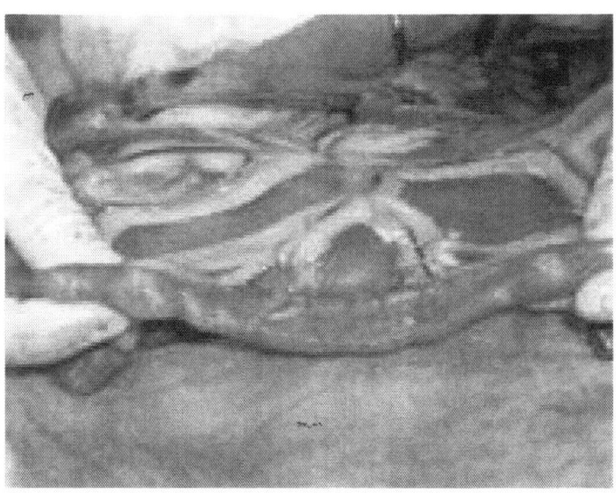

Fig. 7.17 Evaluation of the small intestine during an exploratory.

3. Begin exploration as soon as the linea is incised. Note the appearance of the abdomen, the presence and character of fluid, organ positions, gut motility and

the integrity of the peritoneal wall and diaphragm. Traditionally, the abdomen is explored cranially to caudally, but the order of exploration doesn't really matter as long as you use a systematic approach and do it the same way every time. Look at and feel everything. And be sure to evaluate the entire abdomen before addressing any problems you identify. Good technique and tissue handling throughout the procedure are a must.

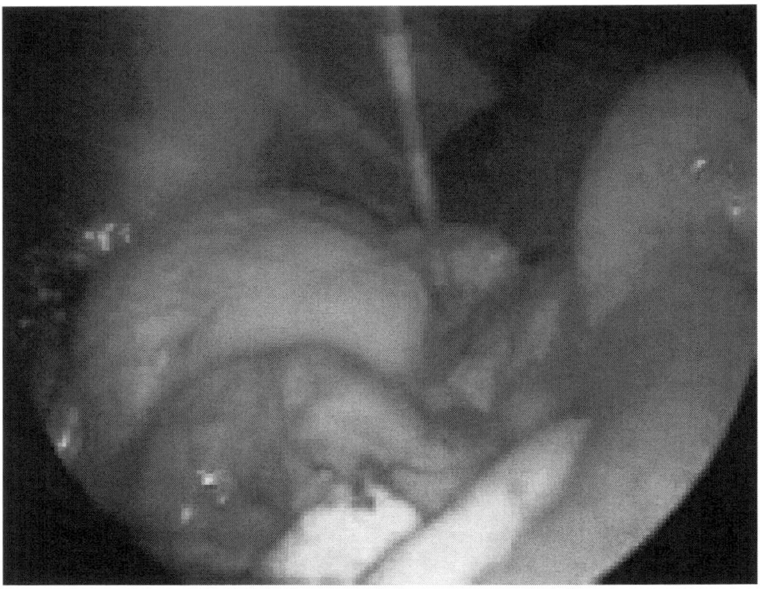

Fig. 7.18 A laparoscopic exploratory evaluating the kidney, intestine, liver and adrenal gland.

4. Complete your plan once the entire abdomen has been evaluated. Isolate the area of interest with moistened laparotomy sponges, particularly if entering a lumen. Use stay sutures, atraumatic retractors, clamps and forceps to minimize tissue damage. Use appropriate suture materials—ideally, in most cases, this involves monofilament, absorbable sutures on small taper needles. Biopsies are indicated for any abnormalities noted (e.g., thickened bowel, lymphadenopathy, lesions or nodules on the liver or spleen). Local lavage of the surgical site or, if warranted, lavage of the entire abdomen with warm, sterile saline solution can be done at the end of the procedure. Evacuate as much fluid as possible from the abdomen, and perform sponge and instrument counts before closure.

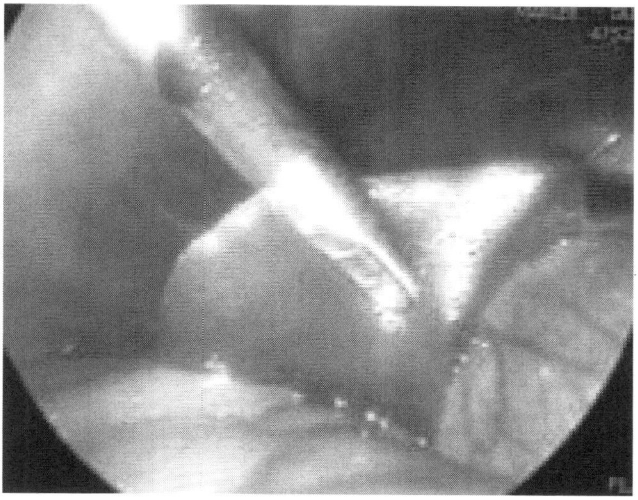

Fig. 7.19 A liver biopsy sample being obtained via laparoscopy.

5. Close the abdomen by using the external fascia of the rectus abdominis, not the muscle or subcutaneous fat. A local block using bupivacaine can be performed along the linea for additional analgesia. Then perform standard subcutaneous and skin closures. To ensure a good recovery, administer appropriate analgesia, heat support and supportive care.

8

GASTROINTESTINAL SURGERY

Gastrointestinal surgery is performed very commonly in small animals for biopsy, removal of a foreign body, upper gastrointestinal bleeding, resection of a nonviable segment of intestine, resection of nonviable portion of the stomach, and resection of a neoplasia. Gastrotomy, gastrectomy, enterotomy, and enterectomy are the procedures routinely performed. General principles common to all of these procedures have to be followed.

Peritonitis is a life-threatening complication that can happen in 13 to 20% of the cases. To reduce the risk of peritonitis, it is important to minimize the trauma to the tissue and to recognize the tissue viability. Peritonitis can develop in the immediate postoperative period because of severe contamination during the surgery or leakage of the suture line. Leakage happens if the tissues are not viable or when the surgical technique was not appropriate. Peritonitis can happen also 3 to 4 days after surgery at the peak of debridement. Dehiscence of an anastomosis happened because the debridement phase goes beyond the suture line. If the tissue handling or the surgical technique were not appropriate, the debridement will be exacerbated and dehiscence can occur.

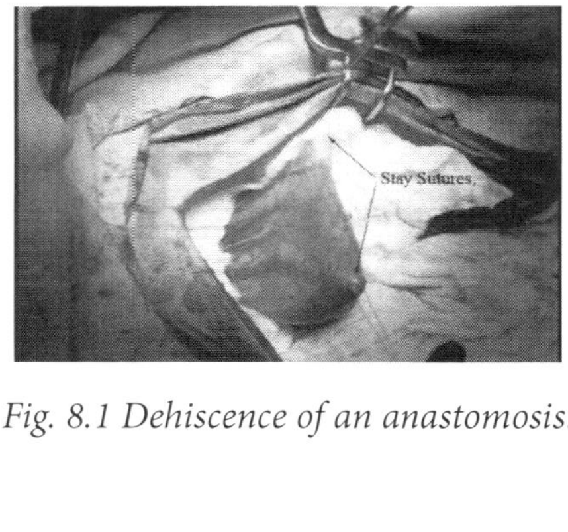

Fig. 8.1 Dehiscence of an anastomosis.

The most common reason for performing a gastrotomy in veterinary medicine is for foreign body retrieval. Through a standard midline approach, the entire gastrointestinal tract should be examined and palpated to determine the location and extent of pathology. If a gastrotomy and enterotomy has to be performed, begin with the gastrotomy (see section on management of foreign bodies). The stomach is exteriorized for gastrotomy. Place stay sutures to facilitate exteriorization and atraumatic manpulation of the stomach. With moist laparotomy sponges, isolate the stomach from the remainder of the abdomen to prevent contamination. Incise the stomach through a relatively avascular area midway between the gastric and gastroepiploic vessels (greater and lesser curvature).

ANTIBIOTIC PROPHYLAXIS

Gastrointestinal surgery without a peritonitis is considered as a clean-contaminated surgery. Therefore, antibiotics are required during the procedures. Most common bacteria encountered during gastrointestinal surgery are: Escherichia coli, Enterococcus spp., Streptococcus, Staphylococcus aureus, Proteus, and Bacteroides fragilis. In the stomach, the bacteria population is more limited than in the rest of the intestine because of the acidic environment. In the colon, the population of anaerobic bacteria is elevated.

Before surgery, the patient is placed on prophylactic intravenous antibiotic. For the surgery of the stomach, ampicillin or first-generation cephalosporin (20 mg/kg IV every 60 min) can be used. For the rest of the intestine, second-generation antibiotics are preferable, because they have a broad spectrum against gram + and -. Cefoxitin (20 mg/kg IV every 90 min) is the antibiotic of choice. A combination of ampicillin enrofloxacin can be used also especially if a peritonitis is present. Antibiotics are interrupted at the end of the procedure unless the animal has peritonitis. The animals need to be watched closely for signs of infection.

CHOICE OF SUTURE MATERIAL AND NEEDLES

A wide range of suture materials has been used during gastrointestinal surgery. It is recommended to use synthetic absorbable monofilament suture materials. Synthetic absorbable sutures are absorbed primarily by hydrolysis, which is very predictable. Monofilament sutures are less traumatic than braided sutures when they are placed through the wall of the gastrointestinal tract. However, they have more memory which makes them harder to work with. A monofilament suture size 4.0 or 3.0 is recommended for gastrointestinal surgery. A small suture material will more likely trigger less inflammation. Glycomer 631 (Biosyn) or Monocryl is a very good choice since it retains 50% of its tensile strength for 3 weeks. A taper needle is recommended for placement sutures in the stomach wall and the intestine, because it does not cut the fibers of collagen.

SUTURE PATTERN

The gastrointestinal tract is closed with a simple continuous suture pattern to provide apposition of all the layers of the wall of the gastrointestinal tract. It is important to place the suture full thickness to make sure the submucosa is incorporated. This suture technique will provide primary healing of the wall of the gastrointestinal tract. When a simple continuous pattern is used, it is recommended to place the first and the last stitches with knots outside the incision because it may provide a very good seal. When the sutures are placed, it is important to avoid traumatizing the wall of the gastrointestinal tract. It is then better not to grab the wall of the gastrointestinal tract with grasping forceps as much as possible. When the sutures are placed, it is important to follow the curvature of the needle to reduce trauma to the soft tissue.

ASEPSIS

Every effort should be made to prevent contamination of the abdominal cavity during gastrointestinal surgery. The stomach or the loop of intestine is "packed off" the rest of the abdominal cavity. Layers of moist laparotomy sponges are used. Intestinal content is moved away for the surgical site with gentle manipulation, and atraumatic clamps are placed to prevent the intestinal content from coming back. Stay sutures are placed in the stomach to elevate the stomach wall and prevent spillage of stomach content into the peritoneal cavity.

After completion of the surgery, the peritoneal cavity is lavaged with warm sterile saline solution to remove gastrointestinal spillage and blood clots. Usually 1 liter of saline is used for a 10-kg dog. The entire fluid is eliminated with surgical suction to get the peritoneal cavity as dry as possible.

Gloves and instruments are then changed. New surgical towels are placed on the edges of the laparotomy. The abdominal cavity is then closed routinely.

Incision

First, make a stab incision using a scalpel blade and then extend the incision with Metzembaum scissors. When stabbing the stomach, you will notice significant resistance to penetrating the submucosa and mucosa. Do not cauterize or ligate bleeders, as hemorrhage will stop when you suture the stomach closed. If suction is available, suction the stomach contents and prevent spillage. Using an instrument, explore the gastric cavity and retrieve foreign bodies. Adding a third and fourth stay suture on either side of the incision may facilitate visualization. As much as possible, do not contaminate your gloves or place your gloved fingers into the stomach. Enlarge the size of the incision if necessary to remove the foreign body.

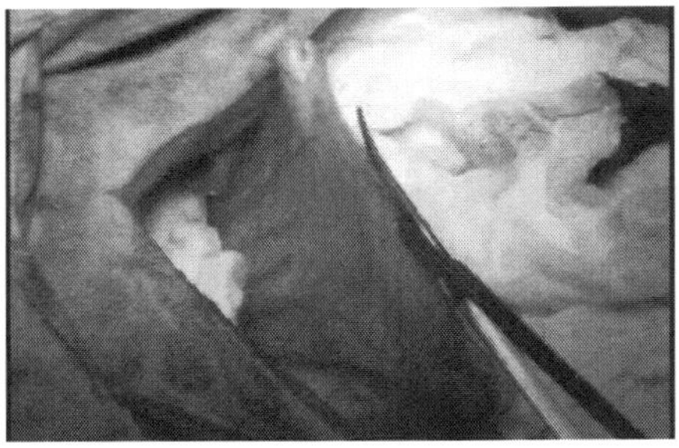

Fig 8.2 a Gastric Incision

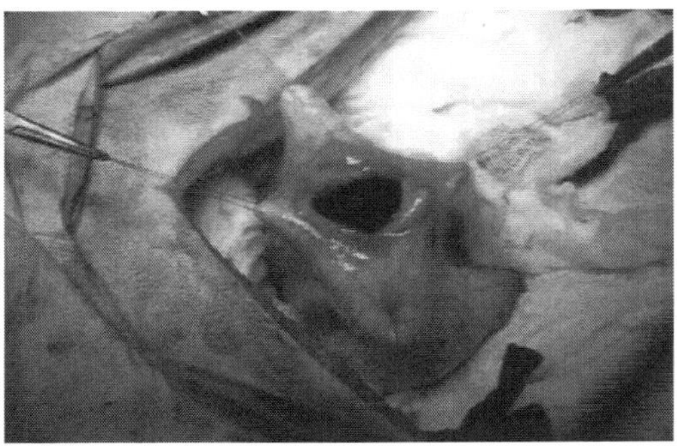

Fig. 8.2 b Additional Stay Sutures

ASSESSMENT OF VIABILITY

Stomach

No objective criteria exist to evaluate the gastric wall viability. Absence of peristaltic wave, pale greenish to gray serosal color, thin gastric wall, and lack of bleeding after partial-thickness incision are signs of gastric wall devitalisation.

Intestine

Viability is assessed by coloration of the serosa, peristalsis, pulse in the jejunal arteries, and utilization of intravenous fluorescein. Fluorescein at the dose of 15 mg/kg is injected intravenously. Fluorescein emits a gold-green fluorescence when exposed to ultraviolet light. Viable intestine has a smooth uniform green-gold fluorescence. Hyperemic intestine has a brighter color than normal. Nonviable intestine has patchy fluorescence or no fluorescence. The fluorescence can also be located around blood vessels.

SURGERY OF THE STOMACH

Gastrotomy

Gastrotomy incision is performed in a relatively avascular area, midway between the lesser and the greater curvature of the stomach. One stay suture is placed at each end of the planed gastrotomy incision. Stay sutures are used to manipulate the stomach and bring the stomach wall to the edge of the abdominal incision. The stomach is walled off with moist laparotomy sponges to prevent contamination of the abdominal cavity with gastric content. An 11 blade is used to stab the gastric wall next to one stay suture. Then, a Metzenbaum scissors is used to extend the gastrotomy incision. The incision should be long enough to remove the foreign body without tearing the stomach wall. Hemorrhage occurs mainly in the seromuscular layer and can be controlled by ligation with fine absorbable sutures. The foreign body is extracted and the gastrotomy closed. Gastrotomy is closed with one-layer simple apposition pattern with 3-0 monofilament absorbable suture. The abdominal cavity is lavaged with warm saline prior to closure. Surgical gloves are changed, and a new pack of instruments is then used to complete the abdominal closure.

Gastrectomy

Gastrectomy is required to resect the nonviable part of the stomach wall during GDV, or resection of an ulcer or a neoplasia. Stomach wall resection for a gastric ulcer requires wide margin since gastric ulceration could be associated with gastric adenocarcinoma or gastric lymphoma.

Two options are available: gastrectomy with either traditional suture technique or stapling suture. Gastrectomy with automatic stapling equipment is associated with the best post-surgical outcome during GDV. This is the recommended technique.

GASTRECTOMY DURING GDV

Branches of the short gastric arteries and left gastro-epiploic artery supplying the area to be resected are ligated. Nonviable gastric wall is resected with Metzenbaum scissors to the level of healthy gastric tissue. Closure with suture material is possible. This technique is associated with 60% mortality. If autostapling equipment is available, a thoracoabdominal device (TA 55 or 90) can be used to perform the gastrectomy (Figure 1). The TA 55 or 90 placed two staggered rows of staples respectively 55 or 90 mm long. The length of staples for stomach wall resection should be 4.8 mm. The TA 55 or 90 is first clamped on healthy stomach wall. The two rows of staples are then fired, and a 15 blade is used to resect the devitalized stomach wall. The TA is then released. The suture line is then inspected to be sure that all the staples are in correct position, and then a simple inverting suture line is applied. The advantages of this technique include decreased surgical time and decreased abdominal contamination from gastric spillage. Mortality rate with the autostapling equipment is close to 10%. Stomach rupture at the time of surgery is associated with severe peritonitis.

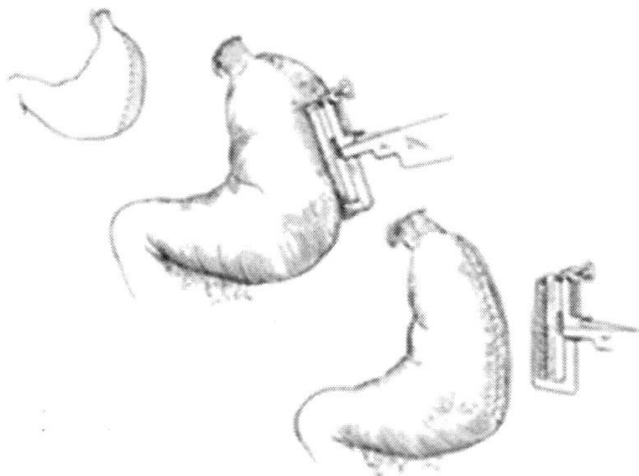

Fig. 8.3 Gastrectomy During GDV.

Gastrectomy for Ulcer or Mass

After identification of the stomach ulcer or tumor, the healthy stomach is retracted with stay sutures to prevent gastric spillage. The stomach wall is then resected around the lesion. The stomach wall is highly vascular. Bleeding blood vessels on the line of incision are ligated with 4.0 monofilament absorbable sutures. Staples can be used to close the gastrectomy. Hand suture can be performed if stapling equipment is not available. Two inverting continuous sutures have been used to close the stomach after gastrectomy. A Cushing suture followed by a Lembert is the traditional technique. However, one simple continuous apposition suture can be used to close the stomach.

SURGERY OF THE INTESTINE

Enterotomy

A variety of foreign bodies can be ingested by young animals. Linear foreign bodies are more common in cats. Once an object has passed through the pylorus, the next smallest lumen are the distal duodenum and the proximal jejunum. Plication of the intestine indicates the presence of a linear foreign body. Usually one end of the linear body is still located in the stomach and needs to be released first.

Particular attention is paid to mesenteric border, especially with a linear foreign material. Linear foreign body can cut through the mesenteric border and induce leakage of intestinal content. Multiple enterotomies may be required to remove the entire linear foreign body. Enterectomy is required if the mesenteric border has been damaged. Enterotomy is performed immediately distal to the foreign material. Intestine is walled off the abdominal cavity to prevent leakage of intestinal content inside the abdominal cavity. The incision on the antimesenteric border should be long enough to extract the foreign body without tearing the intestine. The enterotomy is then closed with a simple interrupted appositional pattern with 4-0 monofilament absorbable suture. Since the submucosa is the only holding layer, it needs to be incorporated in the suture. The submucosa tends to retract away from the edge of the incision. The everted mucosa can be trimmed away to allow better exposure of the submucosa. A serosal patch can be performed to improve blood supply to the enterotomy site and increase its tensile strength. The abdominal cavity is lavaged with warm saline prior to closure. Surgical gloves are changed, and a new pack of instruments is then used to complete the abdominal closure.

Enterectomy

An enterectomy is indicated if the intestine is not viable (intussusception, volvulus), perforated by a linear foreign material at the mesenteric border, or if an intestinal neoplasia is present. End-to-end anastomosis is the preferred technique to perform an enterectomy. Intestine is walled off the abdominal cavity to prevent leakage of intestinal content inside the abdominal cavity. Enterectomy requires double ligation of the branches of the arteriovenous supply of the portion of intestine to resect.

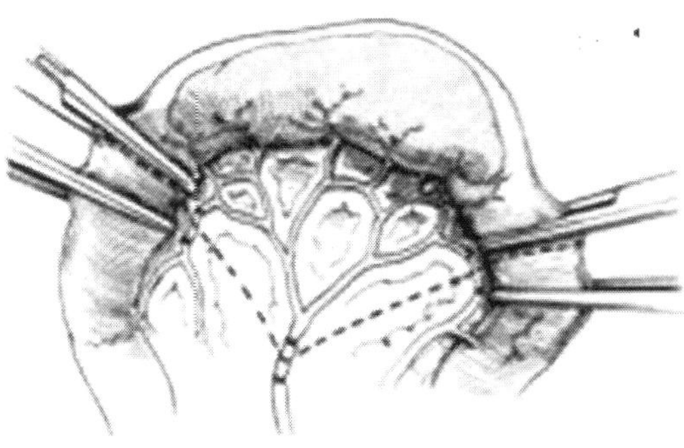

Fig. 8.4 Enterectomy.

The terminal arcade is also double ligated. Transection is performed between the double ligations. Crushing clamps are then placed on the portion of intestine that is going to be resected. The clamps are either placed perpendicular or at a slight angle toward the normal intestine. Non-crushing clamps are placed on the normal intestine 4 to 5 cm away from the enterectomy site after milking away the intestinal content. The intestine is transected with a scalpel blade using the crushing clamps as guide. Since the submucosa is the only holding layer, it needs to be incorporated in the suture. The submucosa tends to retract away from the edge of the incision. The everted mucosa can be trimmed away to allow better exposure of the submucosa.

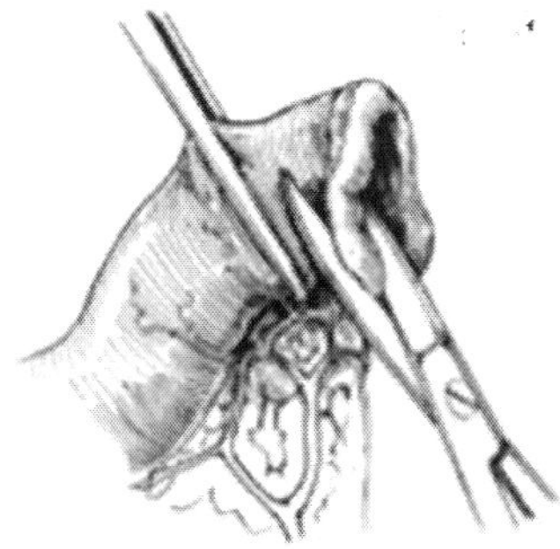

Fig. 8.5 Non-crushing clamps.

Anastomosis is performed with simple interrupted or continuous appositional pattern with 4-0 monofilament absorbable suture.

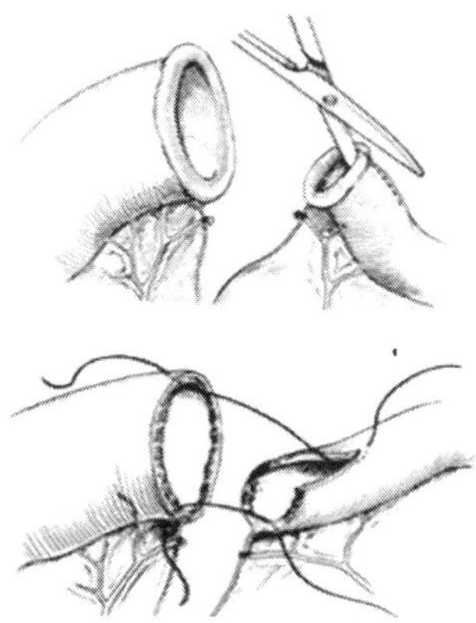

Fig. 8.6 Anastomosis is performed.

If the two extremities of the intestine are not of equivalent diameter, the smaller diameter segment can is incised on the antimesenteric border to create a larger spatulated edge. A serosal patch can be placed around the enterectomy site to improve blood supply and tensile strength. The abdominal cavity is lavaged with warm saline prior to closure. Surgical gloves are changed, and a new pack of instruments is then used to complete the abdominal closure.

POSTOPERATIVE MONITORING

Postoperatively the patients should be monitored for signs of peritonitis, hypotension, disseminated intravascular coagulation, and pain. Patients can be fed 8 hours after surgery unless the patient is vomiting. Feeding tubes might be required to support those patients. Peritonitis occurs in 13 to 20% of the cases. Leakage after gastrointestinal surgery can happen within 24 hours of surgery because of nonviable tissue or inappropriate surgical technique. It can also happen 3 to 4 days after surgery at the peak of the debridement phase. Clinical signs of peritonitis will be tachycardia, tachypnea, acute abdominal pain, hypoglycemia, and hyperthermia. An abdominocentesis is then required to confirm the presence of bacteria in the abdomen. If bacteria are presented past 8 hours after surgery, the anastomosis is leaking and a second surgery is required.

Insulin Resistance: Diagnosis and Management

nsulin resistance is a condition in which a normal amount of insulin produces a suboptimal biologic response. Insulin resistance may result from problems occurring before the interaction of insulin with its receptor (e.g., insulin-binding antibodies), at the receptor (e.g., altered insulin receptor binding affinity or concentration), or at steps distal to the interaction of insulin and its receptor. Post-receptor problems are difficult to differentiate clinically from receptor problems, and both often coexist. In dogs and cats, receptor and post-receptor abnormalities are usually attributable to obesity, inflammation (such as occurs with pancreatitis or gingivitis), a disorder causing excessive secretion of a potentially insulin-antagonistic hormone (such as cortisol in dogs and cats or growth hormone and T4 in cats), or a disorder that causes a deficiency of hormone necessary for insulin action (such as thyroid hormone).

No insulin dose clearly defines insulin resistance. For most diabetic dogs and cats, control of glycemia can usually be attained using 1.0 U or less of NPH, lente insulin or glargine (cats) per kilogram of body weight given twice daily. Insulin resistance should be suspected if control of glycemia is poor despite an insulin dosage in excess of 1.5 U/kg, when excessive amounts of insulin (i.e., insulin dosage > 1.5 U/kg) are necessary to maintain the blood glucose concentration below 300 mg/dL, or when control of glycemia is erratic and insulin requirements are constantly changing in an attempt to maintain control of glycemia. Failure of the blood glucose concentration to decrease below 300 mg/dL during a serial blood glucose curve is suggestive of but not definitive for the presence of insulin resistance. An insulin resistance-type blood glucose curve can also result from stress-induced hyperglycemia (cats), the Somogyi response, and other problems with insulin therapy, and a decrease in the blood glucose concentration below 300 mg/dL can occur with disorders causing relatively mild insulin resistance. Serum fructosamine concentrations are typically greater than 500 µmol/L in animals with insulin resistance and can exceed 700 µmol/L if resistance is severe.

Two diseases that have the potential to cause the most severe insulin resistance are hyperadrenocorticism and hypersomatotropism (acromegaly), although insulin resistance may also be mild or variable. Approximately 80% of cats with hyperadrenocorticism and nearly all cats with hypersomatotropism will develop diabetes mellitus. Hyperadrenocorticism is rare: 75% to 80% of cats have pituitary-dependent disease and 20% to 25% have cortisol secreting adrenocortical tumors. In rare circumstances, adrenocortical tumors secrete other steroid hormones (e.g., progesterone). However, clinical signs are identical to those of hypercortisolism, and diabetes mellitus may develop as well. In addition to PU/PD and weight loss, which are usually due to concurrent diabetes mellitus, typical clinical signs are abdominal enlargement, an unkempt seborrheic hair coat, thinning of the hair coat, failure of hair to regrow, or alopecia and muscle weakness. Severe cases may have thin, fragile skin that tears easily. Cats with large pituitary masses may have CNS disturbances. However, clinical signs may also be mild and hyperadrenocorticism is often not suspected until it becomes evident that the diabetes is difficult to regulate. The dexamethasone suppression test is the preferred screening test. Whether poorly regulated diabetics do indeed have hyperactivity of the hypothalamus-pituitary-adrenal gland axis that leads to abnormal test results is controversial. Based on recent studies, the dexamethasone test (0.1 mg/kg dexamethasone IV with a pre, 4 and 8 hour post) appears to be a suitable part of the diagnostic workup in diabetic cats suspected of having hyperadrenocorticism and should be carried out only after insulin therapy has been instituted for 6–8 weeks to mitigate the effects of poor glycemic control on the HPA axis.

Hypersomatotropism in cats is caused by a growth hormone (GH)-producing tumor (usually an adenoma) in the pars distalis of the pituitary gland. GH has catabolic and anabolic effects; the latter are in part mediated by insulin-like growth factor-1 (IGF-1). The catabolic effects are mainly due to insulin antagonism and are the reason for the diabetes mellitus. The anabolic effects include proliferation of bone, cartilage, soft tissue, and organs resulting in a large body size, broad head and large paws, weight gain, prognathia inferior, respiratory difficulties because of thickening of pharyngeal tissues, degenerative arthropathy, and organomegaly with potential organ dysfunction. Growth of the tumor may lead to signs of CNS disease. As previously mentioned for hyperadrenocorticism, clinical signs may also be very subtle or even absent. Acromegaly has long been considered a rare disorder. However, it was recently suggested that acromegaly occurs more frequently than previously thought and is most likely underdiagnosed. Because the availability of a validated GH assay for cats is inconsistent, diagnosis is usually based on the finding of high IGF-1 concentration. Two important points should be kept in mind. First, circulating IGF-1 is bound to proteins, which must be removed before measurement. Not all assay methods are equally effective, and intra assay inference of binding proteins may lead to false high IGF-1 levels. Therefore, only assays validated for the cat should be used. Second, IGF-1 concentrations are often low in newly diagnosed diabetic cats and increase markedly after initiating insulin therapy. Low IGF-1 levels have also been seen initially in untreated diabetic cats with acromegaly. This observation is explained by the fact that relatively high insulin concentrations are required in the portal vein for the expression and function of GH receptors on hepatocytes, and this mechanism is impaired in insulin-deficient states. IGF-1 is therefore measured 6 to 8 weeks after initiating insulin therapy.

PROBLEMS WITH INSULIN THERAPY

- Inactive insulin
- Improper insulin syringe
- Diluted insulin
- Improper administration technique
- Inadequate dose
- Somogyi response
- Inadequate frequency of insulin administration
- Impaired insulin absorption
- Anti-insulin antibody formation (rare)

CAUSED BY CONCURRENT DISORDER

- Diabetogenic drugs
- Hyperadrenocorticism
- Diestrus (intact female dogs)

Infection, especially of skin, oral cavity and urinary tract

Chronic inflammation, especially pancreatitis and oral cavity

- Severe obesity
- Hyperlipidemia
- Hypothyroidism
- Hyperthyroidism (cat)
- Acromegaly (cat)
- Renal insufficiency
- Liver insufficiency
- Cardiac insufficiency
- Pancreatic exocrine insufficiency
- Neoplasia
- Glucagonoma
- Pheochromocytoma

Many disorders can interfere with the effectiveness of insulin therapy. The most common disorders causing insulin resistance in dogs include severe obesity, use of diabetogenic drugs (glucocorticoids), hyperadrenocorticism, diestrus, chronic pancreatitis, renal insufficiency, oral and urinary tract infections, hyperlipidemia, and anti-insulin antibodies in dogs receiving beef source insulin. Obtaining a complete history and a thorough physical examination are the most important steps in identifying these concurrent disorders. Abnormalities identified on the physical examination may suggest a concurrent insulin-antagonistic disorder or infectious process, which will give the clinician direction in the diagnostic evaluation of the dog. If the history and physical examination are unremarkable, a CBC, serum biochemical analysis, serum progesterone concentration (intact female dog), abdominal ultrasound, and urinalysis with bacterial culture should be obtained to further screen for concurrent illness. Additional tests will be dependent on results of the initial screening tests.

DIAGNOSTIC TESTS TO CONSIDER FOR THE EVALUATION OF INSULIN RESISTANCE IN DIABETIC DOGS AND CATS

Complete blood count, serum chemistry profile, UA and UMIC

- cPLI (pancreatitis)
- TLI (if suspect EPI)

- Adrenal function testing
- Dexamethasone suppression test (cats)
- ACTH stimulation (likely less affected by concurrent diabetes in dogs)
- Thyroid function testing
- TT4
- fT4 (if TT4 is less than 1.5 μg/dl in a dog or between 2.5–4.0 μg/dl in a cat)
- Serum progesterone levels (diestrus in dogs)
- Serum IGF-1 concentrations (cats with suspected acromegaly)
- Fasting triglycerides and cholesterol
- Abdominal ultrasonography (pancreatitis, neoplasia, adrenal masses or enlargement)
- Thoracic radiographs (cardiopulmonary disease, neoplasia)

9

DOG SURGERY

WHAT TO KNOW WHEN DOG SURGERY HAS BEEN RECOMMENDED

At Countryside Veterinary Clinic, we understand that it can be frightening to receive the news that a dog surgery is being recommended. It is important to understand that it is a recommendation that our veterinarians do not take lightly. If one of our veterinarians is recommending a surgical procedure, rest assured that it is with the best interests of your dog in mind. It is important to us that you understand the reasons why a surgical procedure is being recommended and are able to comfortably make the right decisions regarding your dog's health.

Canine surgical procedures fall into two categories where your dog is concerned, elective procedures and those that are urgently necessary.

Most common elective dog surgery procedures include:

- Spaying
- Neutering
- Dental extractions
- Removal of benign growths of the skin

Some common urgent care surgical procedures include:

- Skin lacerations or abscesses
- Intestinal obstruction from a foreign body
- Internal bleeding
- Torn cruciate or ACL ruptures

- Fracture repair
- Malignant skin tumors
- Bladder stones/urethral blockages
- Spleen cancer

We Follow The Highest Standards Of Veterinary Care

At Countryside Veterinary Clinic, we are committed to the highest standards of excellence in veterinary medicine. Surgical protocols include:

- Pre-surgical assessments. Prior to surgery, the veterinary team verifies the specifics of the procedure, completes a physical exam of the dog, and ensures blood tests have been completed, documented, and reviewed by the veterinarian. Among other things, these precautions help determine if your pet is at risk for complications while under general anesthesia.
- Dedicated surgical suites. To prevent post-surgical infections and cross-contamination, surgeries are performed in a room used only for sterile surgical procedures.
- Surgical attire. Staff must wear disposable caps and masks when entering the surgical suite. Anyone involved in the procedure itself must also wear sterile gowns and single-use gloves.
- Sterile packs and equipment. Surgical instruments are carefully cleaned, sterilized, and wrapped prior to each procedure to help prevent infections.

Making The Decision To Proceed With Dog Surgery

The decision to do surgery involves a discussion with the owner about possible complications and all factors to be considered when deciding what is best for your dog. Factors to think about when considering dog surgery include:

- The age and general health of the dog
- Potential complications from the surgery
- Potential outcome if surgery is not done
- Recovery time and post-operative care required by the owner

Although the decision to have your dog undergo surgery is ultimately up to you, our veterinary team will present you with all the facts and possible outcomes to help you make an informed, ethical, and compassionate decision that is in the best interest of both you and your loyal canine friend.

Dog Pre-Surgical Instructions

Dog pre-surgical instructions vary depending on the type of procedure being performed and whether or not the dog surgery is emergency or planned. However, we will provide you with a set of dog pre-surgical instructions that can be used as a general guideline for preoperative preparations:

- Follow your vet's recommendations for feeding and drinking the day before and/or the morning of the surgery.
    - Most surgeries are done on a fasted dog. In general, you will be asked to not feed your dog after midnight the night before the procedure.
    - Most dogs are allowed to drink until the morning of the surgery.
- Be on time for your canine surgery, as most veterinarians schedule surgeries very tightly, and delays potentially threaten the well-being of the tardy dog, as well as the other dogs in line.
- Listen carefully to the post-surgical instructions from your veterinary care team, and call the hospital if you have any questions regarding your dog's post-op care.

Dog Anesthesia

At Countryside Veterinary Clinic, we adhere to very stringent guidelines for administering dog anesthesia before, during, and, if necessary, after surgery. For example, we first do blood work and then depending upon your dog's overall health, other tests to ensure there is not an overt risk of complications from receiving dog anesthesia.

Dog anesthesia is extremely safe when the dog is stabilized before the procedure and all efforts are made to understand the dog's medical condition before surgery. There is always some risk with anesthesia. However, the risk is extremely low when it is performed by a highly qualified veterinarian and surgical team.

Recovery from surgery depends upon the length of the surgery, the age of the dog, and the amount of pain medication required to keep your dog free from any post-operative pain. Some things to be aware of post anesthesia include:

- It is normal for your dog to be groggy or disoriented for a few hours after receiving a general anesthetic.
- Your dog might sleep deeper or longer for 24 hours after receiving dog anesthesia.
- Your dog might be lethargic for 24 hours after anesthesia due to the dulling effects of anesthesia.

- You might need to help your dog balance during feeding and bathroom breaks for the first 24 hours after surgery.
- Consult your veterinarian for any feeding and/or comfort tips they can provide depending on the kind of anesthesia used and what surgical procedure was performed.

Always remember to call us if you have any questions about your dog's recovery.

POST-SURGICAL CARE FOR DOGS

Just like dog pre-surgery instructions, dog surgery recovery protocols and care vary depending on the type of procedure performed and whether or not the surgery was an emergency. However, we will provide you with a set of dog surgery recovery instructions that can be used as a general guideline for postoperative care:

- If you are leaving your dog during surgery, make sure you know when you should return for pickup.
  - For routine procedures, most dogs can go home a few hours after waking up from anesthesia.
  - For advanced or emergency procedures, extended stays of 24 hours or longer may be necessary in order to monitor vital signs and deliver critical care.
- If you did not do so beforehand, make sure to receive and understand all recovery information, including:
  - How to administer medication, food, and water.
  - The changing of bandages, cleaning of stitches, etc.
  - Assisted care tips.
  - Follow-up appointment scheduling.
- At home, allow your dog to recover in a warm, quiet space of his or her choosing (if possible) to increase comfort and reduce stress.
- For the first 24 hours, monitor your dog closely as he or she recovers. Always call if you have any concerns.
- Limit outdoor exposure for at least 24 hours (supervised), and if necessary, assisted bathroom breaks.
  - Consult your veterinarian for more information on the necessity and duration of limited or restricted outdoor exposure.

- Suture care (stitches): Most surgeries will require some sutures. Your veterinary staff will review the after-care with you, which will include keeping your dog from licking the incision.
    - Most dogs will be sent home with an Elizabethan collar to ensure they do not lick or bite the sutures.
    - Monitor the incision for possible signs of infection, including redness or swelling.
- Continue to follow your dog's recovery program until told to alter or discontinue it by your veterinarian.

BIBLIOGRAPHY

- Aitken ML, Patnaik AK. Comparsion of needle-core (trucut) biopsy and surgical biopsy for the diagnosis of cutaneous and subcutaneous masses: A prospective study of 51 cases. JAAHA 2000; 36:153-157.
- Bertman, S. Handbook to Life in Ancient Mesopotamia. Oxford University Press, USA, 2005.
- Bonfanti U, Bussadori C, Zatelli A, et al. Percutaneous fine-needle biopsy of deep thoracic and abdominal masses in dogs and cats. J Small Anim Pract 2004; 45:191-198.
- Conni Lord. "One and the Same? An Investigation into the Connection between Veterinary and Medical Practice in ancient Egypt." Veterinary Medical Practice, 2015, pp. 140-154.
- David, R. Religion and Magic in Ancient Egypt. Penguin Books, 2003.
- Dernell WS, Withrow SJ. Preoperative patient planning and margin evaluation. Clin Tech Small Animal Pract 1998; 13:17-21.
- Dr. Earl Guthrie. "History of Veterinary Medicine." Iowa State University, Volume 2; Issue 1; 1939, pp. 1-6.
- Dr. Saadi F. Al-Samarrai. "Mesopotamian Medicine." Historical Review of Medicine in the Arab World, Online Accessed 27 April 2020, pp. 129-134.
- Ebrey, P. B. The Cambridge Illustrated History of China. Cambridge University Press, 2010.
- Ehrhart N. Principles of tumor biopsy. Clin Tech Small Anim Pract 1998; 13:10-16.

- Einstein R, Sorgente N, Soble LW, et al. The resistance of certain tissues to invasion: Penetrability of explanted tissues by vascularized mesenchyme. Am J Pathol 1973; 73:765.
- Farese JP, Bacon NJ, Liptak JM, Kow K. Introduction to oncology for the general surgeon. In Tobias KM, Johnston SA (eds): Veterinary Surgery: Small Animal. St. Louis: Elsevier, 2012, pp 304-324.
- Harvey, B. K. Daily Life in Ancient Rome. Focus, 2016.
- Hippocrates. Hippocrates' Works. Harvard University Press, 1990.
- Hornblower, S. & Spawforth, A. The Oxford Companion to Classical Civilization. Oxford University Press, 2014.
- Keay, J. India. Grove Press, 2020.
- Kramer, S. N. History Begins at Sumer. University of Pennsylvania Press, 1988.
- Kramer, S. N. The Sumerians: Their History, Culture, and Character. University of Chicago Press, 1971.
- Kriwaczek, P. Babylon: Mesopotamia and the Birth of Civilization. St. Martin's Griffin, 2012.
- Kulke, H. & Rothermund, D. A History of India. Barnes & Noble Books, 1999.
- Liptak JM. The principles of surgical oncology: Diagnosis and staging. Compend Contin Ed Pract Vet 2009; 31(9):E1-E12.
- Liptak JM. The principles of surgical oncology: Surgery and multimodal therapy. Compend Contin Ed Pract Vet 2009; 31(9):E1-E14.
- McChesney SL, Withrow SJ, Gillette EL, et al. Radiotherapy of soft tissue sarcomas in dogs. JAVMA 1989; 194(1):60-63.
- Pormann, P. E. The Cambridge Companion to Hippocrates. Cambridge University Press, 2018.
- Poston GJ. Is there a surgical oncology? Textbook of Surgical Oncology. London: Informa Healthcare, 2007, p 1.
- R. Somvanshi. "Veterinary Medicine and Animal Keeping in Ancient India." Ancient Agri-History, Issue 2; Volume 10; 2006, pp. 1-12.
- Shaw, I. The Oxford History of Ancient Egypt. Oxford University Press, U.S.A., 2004.

- Simpson AM, Ludwig LL, Newman SJ, et al. Evaluation of surgical margins required for complete excision of cutaneous mast cell tumors in dogs. JAVMA 2004; 224:236.
- Soderstrom MJ, Gilson SD. Principles of surgical oncology. Vet Clin North Am Small Anim Prac 1995; 25:97-110.
- Treistman, J. M. The Prehistory of China. Wren Publishing, 1982.
- Withrow SJ. Surgical oncology. In Withrow SJ, MacEwen EG (eds): Small Animal Clinical Oncology, 3rd ed. Philadelphia: Saunders, 2001, pp 70-76.
- Withrow SJ. Surgical oncology. In Withrow SJ, MacEwen EG (eds): Small Animal Clinical Oncology, 3rd ed. Philadelphia: Saunders, 2001, p 70.

INDEX